ANCIENT MEXICO

TOURING NORTH AMERICA

SERIES EDITOR
Anthony R. de Souza, *National Geographic Society*

MANAGING EDITOR
Winfield Swanson, *National Geographic Society*

ANCIENT MEXICO

Aztec, Mixtec, and Maya Landscapes

BY
GEORGE E. STUART
AND
WINFIELD SWANSON

RUTGERS UNIVERSITY PRESS • NEW BRUNSWICK, NEW JERSEY

This book is published in cooperation with the 27th International Geographical Congress, which is the sole sponsor of *Touring North America*. The book has been brought to publication with the generous assistance of a grant from the National Science Foundation/Education and Human Resources, Washington, D.C.

Rutgers University Press
109 Church Street
New Brunswick, New Jersey 08901

The paper used in this book meets the minimum requirements of American National Standard for Information Sciences—Permanence of Paper for Printed Library Materials, ANSI Z39.48-1984.

Library of Congress Cataloging-in-Publication Data

Stuart, George E.
    Ancient Mexico: Aztec, Mixtec, and Maya landscapes / by George E. Stuart and Winfield Swanson.
        p.   cm.—(Touring North America)
    Includes bibliographical references and index.
    ISBN 0-8135-1892-X (cloth)—ISBN 0-8135-1893-8 (pbk.)
    1. Indians of Mexico—Antiquities—Guidebooks.   2. Mexico—Antiquities—Guidebooks.   3. Mexico—Description and travel—Guidebooks.   I. Swanson, Winfield.   II. Title.   III. Series.
F1219.S94   1992
917.204′255D835—dc20   92-9985
                                                                                              CIP

First Edition

Frontispiece: Stucco head from the Temple of the Inscriptions at Palenque, commonly thought to represent Pacal the Great (A.D. 603 to 683). Now on display at the Museo Nacional de Antropología in Mexico City. Photograph by B. A. Stewart, courtesy of the National Geographic Society.

Series design by John Romer

Typeset by Peter Strupp/Princeton Editorial Associates

◬ Contents

△ Foreword

Touring North America is a series of field guides by leading professional authorities under the auspices of the 1992 International Geographical Congress. These meetings of the International Geographical Union (IGU) have convened every four years for over a century. Field guides of the IGU have become established as significant scholarly contributions to the literature of field analysis. Their significance is that they relate field facts to conceptual frameworks.

Unlike the last Congress in the United States in 1952, which had only four field seminars all in the United States, the 1992 IGC entails 13 field guides ranging from the low latitudes of the Caribbean to the polar regions of Canada, and from the prehistoric relics of pre-Columbian Mexico to the contemporary megalopolitan eastern United States. This series also continues the tradition of a transcontinental traverse from the nation's capital to the California coast.

Mexico embraces the richest historic centers on the North American continent. This guide examines the highlights of nearly 4,000 years of cultural history from the Olmec to the present-day survivals, and will include visits to the ruins of Teotihuacan on the central Mexican plateau and the magnificent Maya archaeological sites on the Yucatán Peninsula. George E. Stuart is staff archaeologist of the National Geographic Society, and Winfield Swanson is managing editor of *National Geographic Research Exploration*.

Anthony R. de Souza
BETHESDA, MARYLAND

◿ **Acknowledgments**

We acknowledge the dedicated work of the following cartographic interns at the National Geographic Society, who were responsible for producing the maps that appear in this book: Nikolas H. Huffman, cartographic designer for the 27th IGC; Patrick Gaul, GIS specialist at COMSIS in Sacramento, California; Scott Oglesby, who oversaw production of the shaded relief artwork; Lynda Barker; Michael B. Shirreffs; and Alisa Pengue Solomon. The shaded relief in this volume was drawn by Nikolas H. Huffman. Assistance was provided by the staff at the National Geographic Society, especially the Map Library and Book Collection, the Illustrations Library, the Cartographic Division, Computer Applications, and Typographic Services. Special thanks go to Susie Friedman of Computer Applications for procuring the hardware needed to complete this project on schedule.

We thank Lynda Sterling, publicity manager and executive assistant to Anthony R. de Souza, the series editor; Richard Walker, editorial assistant at the 27th International Geographical Congress; and Tod Sukontarak, who served as photo researcher, and Natalie Jacobus, geography interns at the National Geographic Society. They were major players behind the scenes. Many thanks, also, to all those at Rutgers University Press who had a hand in the making of this book—especially Kenneth Arnold, Marilyn Campbell, Karen Reeds, and Barbara Kopel.

Errors of fact, omission, or interpretation are entirely our responsibility, and any opinions and interpretations are not necessarily those of the 27th International Geographical Congress, which is the sponsor of this field guide and the *Touring North America* series.

PART ONE

Introduction to the Region

Ancient Mexico

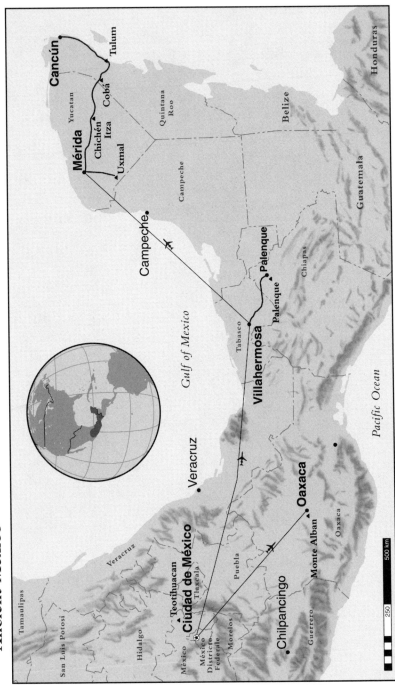

THE REGION

The term Mesoamerica refers to an area defined not by nature or politics, but rather by culture—the customs and ways of people. In the case of Mesoamerica, the boundaries are somewhat vague since they have been established as a kind of average, based on the maximum extent achieved by the ebb and flow of thousands of years of interrelated culture history and movement. Thus defined, Mesoamerica extends from what is now north-central Mexico far into the southeast so as to include the Maya area and beyond, to the Pacific coast of present-day Costa Rica.

Mesoamericans include many familiar cultures of pre-Columbian times—among them the Olmec, Zapotec, Maya, Toltec, Mixtec, and Aztec—and their modern-day descendants. All these shared the use of a complicated calendar, hieroglyphic writing, astronomy, and a complex religion and cosmology. Mesoamerican architects erected masonry palaces and pyramid-temples, while sculptors carved monuments dedicated to political propaganda and the supernatural world. Farmers, a key component of ancient Mesoamerican cultural prosperity after about 2000 B.C., grew maize, beans, and squash, the all-important trio of food crops; other cultigens, including cacao, served as important items of trade throughout the region.

The Mesoamerican landscape is one of incredible variety. It begins in the northern reaches as broad sweeping desert and ends in forested hill country far to the southeast. In between lie lofty snow-capped volcanoes, jumbled mixtures of mountains and valleys, broad humid coastal marshes cut by sluggish rivers, and the great limestone plain of the Yucatán Peninsula.

The whole of this area lies within the tropics, resulting in a yearly cycle marked by the cadence of alternating dry and wet seasons. These vary somewhat in length and in timing, but generally the dry season lasts from October to May. Differences in rainfall over the whole area have created a vegetation cover that ranges from desert to rain forest and includes nearly every possibility in between.

This land of geographical contrasts can conveniently be divided into two broad topographical zones—Highlands and Lowlands—each distinct from the other in terms of climate, vegetation, and natural resources. Together, they form a varied stage upon which people have lived for more than 10,000 years. Throughout that time, the cultural interaction between the Mesoamerican Highlands and Lowlands played a crucial role in the dynamics of human adaptation and the complicated interplay of goods and ideas.

CHRONOLOGICAL PERIODS

Anthropologists, archaeologists, and others who deal with deep cultural time in Mesoamerica customarily divide its culture history into named chronological periods which are useful mainly as convenient frameworks.

The scheme we will use begins with the Paleoindian Period (before 7000 B.C.), and goes on to include the Archaic Period (ca. 7000–2000 B.C.), the Preclassic Period (2000 B.C.–A.D. 250), the Classic Period (A.D. 250–900), and the Postclassic Period (A.D. 900–1521). The historical era includes the Colonial Period (1521–1821) and the Modern Period (1821 on).

PALEOINDIAN HUNTERS (? TO 7000 B.C.)

We do not yet know with any certainty exactly when the human story of Mesoamerica began. Archaeological evidence, which is both indirect and highly controversial, has been uncovered at a handful of sites in North and South America suggesting that the date of the initial peopling of Mesoamerica may someday be pushed back to 30,000 years ago or even earlier. In fact, the question of when people first came to the Americas remains one of the fundamental unsolved problems of American archaeology. (The ultimate solution of that problem lies in finding an indisputable human record predating the Clovis time horizon of 11,500 years ago.)

Whatever the final conclusion of the quest for the earliest Americans turns out to be, evidence of the apparently earliest human presence in Mesoamerica lies in isolated surface finds of Clovis projectile points—a hallmark of the Paleoindians, or Ice Age big-game hunters who lived in many regions of the Americas beginning about 11,500 years ago. Surface finds of Clovis points range from northern Mexico to highland Guatemala, but to date none have been excavated in archaeological context. For example, a Clovis-type fluted point was found near San Juan Guelavía in the Oaxaca Valley, and broken and burned animal bones of Late Pleistocene times have been found in nearby Cueva Blanca. (Kent V. Flannery provides an excellent summary of the situation in *The Cloud People*.) As a result, we must look at a slightly later period for solid evidence of Paleoindian hunters in the area.

That evidence consists of weapon points—and those are not of the Clovis type—found in direct association with the bones of extinct mammoths at the site of Santa Isabel Iztapán, near Mexico City. These appear to date to somewhere between 10,000 and 9,000 years ago, and were found in the geological layer known as the Becerra Formation. That same stratum yielded two other important remains of this early period—the famed skeletal remains of Tepexpan Man (actually a female), found nearby, and, perhaps, the remarkable Tequixquiac carving, the sacrum of an extinct form of llama slightly modified to make it look like an animal head. ("Perhaps" because the Tequixquiac llama sacrum may be older than both the Santa Isabel Iztapán and Tepexpan finds. Stone tools found with it suggest Tequixquiac as a candidate for a pre-Clovis, or "pre-projectile point" time horizon. In this respect, then, Tequixquiac may fall into the category of putatively very early—and highly debatable—discoveries at Valsequillo, near Puebla, Mexico, and other sites.) Stone projecticle points apparently dating to about the same period as the Iztapán and Tepexpan finds have been found in the lowest levels of caves in Tamaulipas state, to the north, and also in the Tehuacán Valley of Puebla state.

All finds of the Paleoindian Period in Mesoamerica together yield but a vague and shadowy picture of a widespread but relatively small population of big-game hunters and gatherers of wild plant foods, who were probably organized as small bands, and moved with the seasons to take full advantage of food procurement—a general picture that holds true for many other areas of the Americas. With the end of the last Ice Age, however, came the extinction of the mammoth and other large animals, forcing a fundamental change in the patterns of human life and survival by about 9,000 years ago—or, to switch to absolute dates, 7000 B.C. That change, which doubtless seemed gradual to the generations who lived through it, nonetheless defines the end of the Paleoindian Period and the beginning of the Archaic Period.

ARCHAIC PERIOD GATHERERS AND PLANTERS (7000–2000 B.C.)

So far, ancient Mesoamericans have appeared only in terms of sporadic finds of Paleoindian artifacts and extremely rare hunting camps or "kill" sites of that remote period. With the more widespread and detailed evidence pertaining to the Archaic Period, the people themselves begin to take more tangible form in our chronological sketch of culture and change.

Throughout the 5,000-year span between 7000 and 2000 B.C., and in virtually all areas of the Mesoamerican Highlands and Lowlands, the archaeological record reveals an ever-changing pattern of successful adaptation to a wide variety of environmental niches. The old hunting tradition persisted—indeed, it does to this day—but in the wake of the extinction of the Ice Age megafauna the nature of the quarry had changed to deer and smaller animals. And, as Barbara Voorhies (University of California at Santa Barbara) has revealed at Isla Chantuto and nearby sites on the Pacific coast of Chiapas, many groups of the period were quite successful in exploiting shellfish, shrimp, and other marine life of the shoreline lagoons and estuaries. At the same time and later, the presence of metates (milling stones) at sites in the interior suggests an increasing emphasis on the gathering of wild plants, nuts, and seeds—a pattern of subsistence that led directly to the invention of agriculture in Mesoamerica.

The main records of the all-important development of cultivated food plants in the area, mainly maize, beans, and squash, appear in two places—the dry caves of Tamaulipas and in the arid Tehuacán Valley of Puebla. Richard S. MacNeish of the Andover Foundation for Archaeology, Massachusetts, has unraveled the nature of the process. His main conclusions are that Mesoamerican agriculture developed slowly, in multiple centers, and that the three main kinds of plant staples came under domestication in different sequences. In general, the process appears to have been a gradual one, punctuated by episodes of accident and invention, and it took place between about 5000 and 2500 B.C. By the end of the Archaic Period, Mesoamerica was essentially a farmers' world.

Another notable development of the Archaic Period lies in the invention of pottery. Wide-mouthed, baked clay jars, and other vessels that imitate earlier stone counterparts appear in the strata of around 2300 B.C. in the Tehuacán Valley. And the invention appears to have spread rapidly to, or developed independently in, other regions of Mesoamerica in the centuries that followed.

PRECLASSIC MESOAMERICA (2000 B.C.–A.D. 250)

Ocós Culture

At the beginning of the Preclassic Period, most of Mesoamerica was firmly based on maize agriculture, but with continuing hunting and fishing and, in certain shoreline zones, the efficient exploitation of the resources of the lagoons and sea. The stability which that afforded varied in degree from place to place, and was tempered by the ever-present threats of the tropical world—seasonal rainfall and, more often than not, a delicate balance between prosperity and natural disaster. While the peoples of the time apparently inhabited every corner of the land, and spoke a variety of languages, they nonetheless shared certain traits which we see in the remains of the period. These include the ubiquitous baked clay figurines, perhaps household fetishes connected with fertility; an increasing use of ceramics after about 2000 B.C.; and the development of dwelling types, usually of pole and thatch, which proved perfectly suitable for the environment of both the Highlands and the Lowlands.

Many of the remains of the period cluster into an assemblage of artifacts that archaeologists call Ocós, after a site near the Pacific coast where Mexico and Guatemala meet. The archaeological record of Ocós culture from present-day central Mexico to El Salvador indicates a general homogeneity over a huge area for many centuries during Early Preclassic times. The presence of similar pottery, figurines, and house remains provide but a glimpse of a world of small villages and egalitarian societies—a cultural setting that witnessed an intensification of the interchange of goods and ideas that characterized Mesoamerican culture for the next two and a half millennia.

The Olmec

In this milieu of ubiquitous Ocós presence, the culture we now call Olmec came into being. The term Olmec is a misnomer, but it's too late to change it, and it doesn't matter anyway. It comes from the name Olmeca, literally "the rubber people," who lived in the present-day southern Veracruz area of Mexico during the time of the Spanish Conquest. When stone monuments began coming to notice in the late nineteenth century, they were attributed to these historic peoples. Later, beginning in the 1930s, Matthew W. Stirling, supported by National Geographic Society funding, excavated extensively at Olmec sites and recovered data that led him to attribute a very early date to the archaeological culture. In the controversy, Stirling and the artist Miguel Covarrubias held fast and were eventually proved right in their belief in the chronological precedence of Olmec culture. Stirling and, by extension, the National Geographic Society, are generally credited for the discovery of the Olmec, and the bringing of their incredible culture to both public and scientific notice.

We don't know how or why, or even exactly where it came into being, but whatever those details, Olmec culture—some have even labeled it civilization—shone with a bright light throughout the Middle Preclassic Period (whose span it defines), and contained most of the characteristic traits by means of which we define Mesoamerican culture. Among these we see monumental sculpture, complex religious iconography, the ritual ball game, the wide-reaching trade in exotic materials, and perhaps the first hints of the use of hieroglyphic writing (though the last is debatable).

Some Olmec remains are clearly luxury items of exquisite workmanship which range from statuettes of jade and other green stone to monuments of heavy basalt—altars; figures of animals, people, and supernatural blends of the two; and, of course, the famed colossal heads. Olmec art, one of the first great definable styles of the Americas, reflects a shadowy religion permeated by shamanism and images of the caiman, the jaguar, and the were-jaguar, and politics dominated by individual rulers who are occasionally portrayed in the art.

Of the culture itself, we know very little. The size of sites and their distribution suggest a hierarchical society dominated by a wealthy elite, a series of chiefdoms supported by numerous agricultural hamlets and ruled by lineage heads with the power to commission political and religious art.

The Olmec heartland has traditionally been considered as that area now embraced by the adjacent areas of the Mexican states of Veracruz and Tabasco along the Gulf Coast—a lowland zone of marshes and sluggish rivers. The main sites known here include La Venta and San Lorenzo, which Matthew W. Stirling brought to light in the 1930s and 1940s.

The 1985 discovery of the site of Teopantecuanitlan (or Copalillo) in the Balsas Valley of Guerrero state may ultimately prove to be of great importance. This was apparently a major permanent Olmec settlement and its investigation may extend Olmec chronology back to ca. 1500 B.C. If so, our notion of the extent or location of the Olmec heartland may have to change. The site of Teopantecuanitlan is marked by monumental sculpture, including bas-reliefs in Olmec style which may have been part of a large public building. It is under investigation by archaeologist Guadalupe Martínez Donjuan of Mexico's National Institute of Anthropology and History.

Among more recent investigations at Olmec sites, two are of more than routine interest, for they seek data related to ordinary people (versus the elite) of the Olmec. At La Venta, excavations have revealed dwelling areas. And at the site of El Manatí, Veracruz, Ponciano Ortiz and Carmen Rodríguez, of the Veracruz office of Mexico's National Institute of Anthropology and History, are uncovering not only great numbers of polished stone axes, pottery, and figurine heads, but also remarkable Olmec objects of wood and other perishable material—large sculptured busts as high as a meter (three and a quarter feet), a staff tipped by a shark tooth, and even rubber balls!

Olmec or Olmec-like remains dot much of southern and southeastern Mesoamerica. These range from the reliefs at Chalcatzingo, Morelos, Mexico, to similar remains in the area of Chalchuapa, El Salvador. And in between lie other important Olmec vestiges, including the (now-stolen) bas-relief at Xoc, and other Olmec-related remains at sites along the Pacific slope of the Chiapas–Guatemala zone.

It is easy to get the impression that Middle Preclassic Mesoamerica was completely dominated by the Olmec, but this belief is becoming increasingly tempered by the realization that the Olmec, despite spectacular remains that dominate—indeed, even distort— the physical-archaeological record, were but one of many closely related ethnolinguistic groups who occupied both the Highlands and the Lowlands. In short, the role of Olmec culture in Middle and Late Preclassic times has yet to be clearly defined. And finer tuning is badly needed for the definition of the Olmec art style and its relation to Olmec (and other?) cultures of Preclassic Mesoamerica.

Some archaeologists see Olmec as a parent culture in Meso-america, while others characterize it as a sort of sibling culture—an equal-but-influential partner among many talented Preclassic groups, each a product of its own setting and culture history. Crucial to the resolution of the problem, of course, is the matter of dating Olmec remains. At this point, the consensus among Meso-americanists places the culture between around 1200 and 400 B.C. Some investigators, however, question the evidence for this placement, and advocate a later date which, of course, would completely modify existing views on both the area of origin (Pacific slope versus Gulf Coast, or Guerrero state versus Gulf Coast).

As for Olmec art, some see its distribution as a reflection of influence from the Gulf Coast core area, while others suggest that it indicates the spread of certain religious beliefs among many different cultures.

Among other evidence of extraordinary accomplishment during the Middle Preclassic, we have the monumental sculptures of Monte Albán, Dainzú, and San José Mogote in Mexico's Oaxaca state, some of which reflect the earliest developments toward true writing in Mesoamerica, by about 600 B.C. The origin of Meso-american writing and the nature of its development are still subjects open to debate. In fact, even an adequate definition of writing is elusive. It helps if we define writing loosely—such as, a medium of communication using pictorial or even arbitrary symbols on whatever medium—and avoid the obsolete Eurocentric idea that writing progresses from the pictographic to the alphabetic, or that it must replicate spoken language. Mesoamerican writing systems are classed as hieroglyphic—a mixture of pictures and symbols for either words or sounds (that is, syllables).

Sometimes, massive caches are found. All these things have been referred to as either Olmec or Olmecoid, but in truth the nature of the connection, if any, remains obscure. The extraordinary cache of objects of jade and other luxury items found near Peto, Yucatán, showed up for sale in New Orleans in 1984, and at the town of Chacsinkin, adjacent to the unnamed site of discovery. Through the efforts of W. W. Andrews V of Tulane University, the pieces were returned to Mexico. They are clearly of late Middle

Maya hieroglyphics. The glyphic inscriptions reveal names, places, and dates that were important to the Maya. Photograph by Otis Imboden, courtesy of the National Geographic Society.

Preclassic manufacture, and strikingly Olmec-like, although Andrews believes they should be considered more broadly as general representations of elite art of the time.

That Olmec objects were treasured as heirlooms among the Maya and others is reflected in the discovery of an Olmec jade face deposited in a Late Classic burial found by William Rathje and Jeremy Sabloff at the ruins of San Gervasio, on Cozumel Island.

Late Preclassic Mesoamerica: After the Olmec

Whatever the reality of the Olmec phenomenon, it appears to have died out around 400 B.C., based upon the archaeological sequence at La Venta, the latest pure Olmec site to have been excavated. Further investigation at sites now unknown or unexplored may very well modify our views, not only on the ending date, but also on the nature of the relationship among the Olmec manifestation, contemporary Preclassic peoples, and later Mesoamericans. Archaeologists generally use this point to mark the end of the Middle Preclassic Period and the beginning of the Late Preclassic. The Late Preclassic (ca. 400 B.C. to A.D. 250) witnessed the heyday of a number of differing cultures in the various regions of Mesoamerica.

The round, stepped "pyramid" at Cuicuilco, on the outskirts of present-day Mexico City, dominated the great lake there until the lava flow from the eruption of the volcano Xitle enveloped the site around 200 B.C., giving the then-small center of Teotihuacan, to the north, some impetus in the beginnings of its own greatness as a center of population and power.

Meanwhile, development continued steadily in Oaxaca, where monuments at the major hilltop center of Monte Albán depict what appear to be dead and mutilated captives, some accompanied by what appear to be name glyphs. Other stones of the same date—around 600 B.C.—reflect continuing refinement of the art of writing and the use of the distinctive Mesoamerican calendar.

Izapa, near the Pacific Coast at the present Mexico–Guatemala border, and noted for its sophisticated stelae, gives its name to the greater Izapan area, a huge zone between the Oaxaca–Veracruz axis and the Maya frontier where important developments in sculptural art and the beginnings of hicroglyphic writing took place in Preclassic times, and culminated in the Late Preclassic, after the Olmec demise. The stelae at Izapa depict scenes that appear to be episodes of mythology, probably related to the power of local rulership, and a few monuments bear hieroglyphs.

At nearby Abaj Takalik, on Guatemala's Pacific slope, the Late Preclassic inhabitants carved monuments bearing Long-Count dates—stelae which at once seem to be Izapan (like, say, those of Tres Zapotes or La Mojarra), and also remind archaeologists of Classic Maya monuments, except that they are too early and in the wrong place.

One of the most important of the Late Preclassic polities grew from the center of Kaminaljuyu in the Maya Highlands. Now, these ruins consist of a series of unspectacular dirt mounds on the outskirts of Guatemala City, interspersed with an occasional eroded sculpture or the exposed facade of an adobe platform. The role of Kaminaljuyu in the power politics of the time, however, cannot be overestimated. Apparently the place controlling a key obsidian quarry area and, perhaps, several crucial trade routes, it served as the focus for many of the important developments of the age (though, as always, the pattern of archaeological investigation itself plays a role in this reconstruction). Sophisticated hieroglyphic texts (in too small a sample for decipherment) appear amid the delicate tracery of the Izapan-style depictions of rulers or mythical heroes on the fragmentary masterpiece of Stela 10, and massive tombs stuffed with ceramics reflect a powerful local elite.

The Late Preclassic–Early Classic transition seems to have been a time of crystallizing polities and the beginnings of great concern with status and (on the part of the ruling elite) with the quality of real ancestors and the invention of mythical ones. Trade and market systems thrived, for the goods of the various regions differed in kind and in desirability. The green obsidian from the neighborhood of Teotihuacan was prized by all Mesoamericans. So was the obsidian from the quarries of the Maya Highlands, the jade and jadeite from stream deposits, and the volcanic stone used for the manufacture of metates, or corn grinders. The north shore of Yucatán held vital salt, as did the area of the southwestern Petén, while the rain forests yielded feathers of the quetzal and the macaw, and the all-important plantations of cacao began to thrive on the Soconusco coast in the vicinity of Izapa, and in present-day Belize as well.

As in all regions at all times of human history, the flow of goods was affected by all sorts of local conditions ranging from natural factors of river and mountain barriers to the caprice of changing human ways—enmity and wars; competition and the desire for control; and the inevitable accidents and actions of protagonists whom we shall never know. And with the goods went ideas: Jaguars make an excellent symbol of power. So does an elevated platform as a place to live. Mountains are holy. Caves lead to the underworld and nine is the number. The universe is square and its directions have colors. Ancestors came from a tree in the center of the world. The stars are dead ancestors and so are the days that pass.

The Late Preclassic Period also witnessed the first widespread use of the method of recording dates on monumental sculpture by means of the Long Count, a method employing a five-place numerical notation to record the number of days elapsed since the beginning of the calendar count. At Chiapa de Corzo, Stela 2 bears a date equatable to 36 B.C.—the earliest known using the Long-Count system. Four years later, the remarkable Stela C was set up at Tres Zapotes, Veracruz. A few other known monuments for the succeeding two centuries occur at El Baúl and Abaj Takalik, Guatemala, and their dates cluster in the first two centuries of the Christian Era.

One of the most remarkable and intriguing monuments of Late Preclassic times is the great eight-foot (two-and-a-half-meter) Stela 1 found in 1986 at La Mojarra, near Tres Zapotes. It was found quite by accident by men putting in a dock on the Acula River, beside the earthen mounds of the site. The stone was about six and a half feet (two meters) under water, firmly stuck in the mud of the

LONG-COUNT SYSTEM

The Long-Count system of recording the passage of days customarily relies on a five-place notational system of numbers with respective "place values," most often from bottom to top, of 1, 20, 360, 7,200, and 144,000. Placing numbers in these positions—numbers that are, in effect, multipliers—provides a set of products whose sum is the number of days elapsed since the "base date" of the calendar. Thus, all Long-Count dates are accurate relative to one another but, since the system fell out of use long before our time (its practice in the Maya Lowlands defines the Maya Classic Period), the matching of such Mesoamerican dates with Julian or Gregorian days depends on the correlation used. When Matthew W. Stirling discovered the crucial part of Stela C at Tres Zapotes, he equated the date to 291 B.C. Since that time (1939), the correlation has changed so that the very same date is equatable to 32 B.C. (though you will occasionally see 31 B.C. because archaeologists forget to add the necessary civil year involved in the extrapolation of Gregorian dates back beyond the A.D.–B.C. line). In brief, we used to use the Spinden Correlation; now we use the Goodman–Martínez Hernández–Thompson formula, which puts all dates about 260 years later. This makes the base date of the calendar—for convenience, let's call it 0.0.0.0.0 in the Long Count—equal to 12 August 3114 B.C. From there on, the system is analogous to the workings of an odometer: The day after—our 13 August of that year—would be 0.0.0.0.1; the Maya Classic Period lies between 8.12.14.8.15. (the date on the back of Tikal Stela 29, equatable to A.D. 292 since it is [8 × 144,000] + [12 × 7,200] + [14 × 360] + [8 × 20] + [15 × 1], 1,233,615 days after our August 3114 B.C. date) and 10.4.0.0.0 (a date recorded at Tonina, equatable to A.D. 909).

The Maya stopped using the Long-Count method after 10.4.0.0.0. If they had continued to use it, the next cycle of the count, which the Maya would have put at 13.0.0.0.0, will fall (according to the correlation we are using) on 23 December 2012. This date was a key factor in the harmonic convergence movement of a few years ago which, although laudable, was founded on a random and generally incorrect application of ancient Mesoamerican calendar arithmetic.

riverbank. In early 1987 it was moved to the Museo de Antropología in Xalapa, Veracruz, where it still lies in storage. The four-ton slab of basalt features the low-relief image of an elegantly attired man standing in profile. The major portion of the stela contains a hieroglyphic text of more than 500 glyphs, including two Long-Count dates equatable to 21 May A.D. 143, and 13 July A.D. 156. Future investigations at La Mojarra should do much to clarify the picture of cultural dynamics—and the geographical aspects—of the history of writing in Mesoamerica during Late Preclassic times and later.

And so the practical merged with the esoteric. By the end of Preclassic times, the necessity to manage water and trade (but, above all, water, water, water!), blended with the needs of local rulers seeking divine legitimation, had helped to bring about the formation of Mesoamerican political states in numerous variations from the volcanic plateau of central Mexico to the Lowland rain forest in the heart of the Yucatán Peninsula. Authorities are sometimes inconsistent regarding a precise definition of such things as state and civilization, and on the assignment of those degrees of cultural–political development to a particular people. Thus, we see these Preclassic entities—whatever they actually were—denoted as chiefdoms, as states, as civilizations, or whatever. For our purposes, we have chosen to be rather loose with these definitions, recognizing that in most cases the ideal amount of archaeological evidence for comparison or general ranking has proven elusive.

The Maya Area

The term Maya refers to those people of present-day southeastern Mexico and upper Central America who speak any of the two dozen or so languages of the Mayan family of languages and, by extension, to the archaeological remains of the region. There is some ambiguity in the literature regarding the use of the terms Maya and Mayan. In general, our style is to adhere to Maya as the designation for the people and as the adjectival form as applied to sites and artifacts. By this rule, the only appropriate time to employ the term Mayan is in specific reference to the language family. This point of style avoids the plural forms such as Mayas and, of course, Mayans, both of which are commonly seen in the popular literature.

Specifically, the Maya area may be defined as the Yucatán Peninsula and its base, a region that embraces all of Guatemala and Belize; the Mexican states Yucatán, Campeche, Quintana Roo, most of Chiapas, and part of Tabasco; and the western reaches of Honduras and El Salvador. One often sees the term "the Yucatán" in reference to the Maya area or to its northern part. This relatively recent invention first appeared in print in the 1950s. In the opinion of purists, it is not yet completely accepted, though it is evolving rapidly toward legitimacy through sheer frequency of use. Here we use the simple designation Yucatán for both the state and the extreme northern lowlands, and the Yucatán Peninsula for the physical feature.

In the strictest and most accurate definition of the Maya area, the narrow Pacific coastal plain of Chiapas and Guatemala is best eliminated—a fine point that stems from the facts that the cultural makeup of this strip in pre-Columbian times is both confused and that Mayanists and Mesoamericanists know little about it.

Another area of Mexico where Mayan is spoken lies outside and well away from the area defined by the Yucatán Peninsula. This is the Huastec (or Huaxtec) area of northern Veracruz and southern Tamaulipas states, whose peoples are more culturally affiliated with Mexico's central highlands, and whose pre-Columbian works may include the famed ruins of El Tajín, Veracruz.

The Maya area—roughly half the size of Texas—contains both highlands and lowlands. The Maya Highlands make up the southern portion of the whole, except for the narrow Pacific coastal plain of Chiapas and Guatemala. The Maya Lowlands, to the north, contain southern (Petén) and northern (Yucatán) sub-areas. The whole of the area, from present-day central Guatemala to the northern shore of the peninsula and from the Usumacinta Valley to the Caribbean, was in pre-Columbian times the principal hearth of the spectacular civilization that flourished between about A.D. 250 and 900. Thus, when we speak of Maya civilization, we really mean Lowland Maya civilization.

Most of the southern Maya Lowlands are mantled in rain forest which achieves a general canopy height of about 120 feet (40 meters). Other areas contain expanses of savanna. Toward the north and west, the amount of rainfall lessens dramatically, creating a sort of scrub thorn forest that characterizes the area around Mérida, capital of the Mexican state of Yucatán.

The Lowland Maya version of Mesoamerican culture began in a Late Preclassic to Classic cultural cauldron that blended high politics, the scribal arts, and elite lineage groups powerful enough to commission great monuments and buildings. The process by which all this happened is sketchily understood from the investigation of such sites as Cuello, El Mirador, Nakbe, Cerros, and Uaxactun in the southern Lowlands, and at Loltun Cave, Maní, Dzibilchaltún, and other places in Yucatán.

Norman Hammond's work at Cuello gives us one of the best stratigraphic records of an early Maya place, at least in northern Belize. Briefly, the inhabitants of Cuello were intensively growing corn and using ceramics from around 1500 B.C., or during the Middle Preclassic Period, and they were part of a widespread trade network. In the Late Preclassic Period, they developed special elite architecture and erected a stela in front of a small pyramid. The monument was plain, but it reflected the beginnings of the "stela cult," one of the distinguishing traits of the Classic Maya. Major ceremonial activity ceased at Cuello toward the end of the Late Preclassic with the deposit of a Chicanel ceramic cache atop the pyramid.

CHRONOLOGY

The basis for Maya archaeological chronology lies mainly in the ceramic sequences archaeologists have developed, and each site that has been under lengthy excavation has its own. Each such sequence derives, of course, from the principle of stratigraphy: In any deposit made over a long period, that which is earliest is lowest, and that which is latest is highest. Thus, whenever any archaeologist digs a place, one of the first tasks at hand is to establish a basic chronology, and since pottery not only reflects the most subtle changes of ancient fad and fashion, but is incredibly durable as well, it is the chief raw material of the endeavor. By digging in arbitrary levels and keeping track of the material that comes out, the archaeologist classifies ceramics by shape, temper, and decoration into types and varieties. These, in turn, form sets that by degree of difference are separable from those above and those below in the stratigraphy. Each such set, therefore, defines a chronological phase, and is given a name. The name may be anything for, like other period designations, it is but a convenience. Thus, the University of Pennsylvania archaeologists employed Yucatec day names (Cimi, Manik, Eb, etc.) for the ceramic periods at Tikal. Whatever one uses, reference is always made to the sequence established by Robert E. Smith for the site of Uaxactun, and published by the Middle American Research Institute at Tulane University in 1955. Smith's system still works well for the Petén region and, besides, it was the first successful of all such attempts. For his phases, Smith used the names of gods in the Popol Vuh, the famous "national book" of the Quiché Maya. In his system, the phases Mamon and Chicanel are made up of wares on the Late Preclassic Period; and Tzakol (actually Tzakol 1, 2, and 3) and Tepeu (1 and 2) define the Classic Period.

Current work at Nakbe, Guatemala, by Richard Hansen, and the efforts of David Freidel at Yaxuna, Yucatán, may push some Lowland Maya cultural developments back in time, but the data from both projects is still in the process of being collected and analyzed. Among the finds at Nakbe are possible Chicanel tombs with painted hieroglyphic texts.

Earlier excavations by Freidel at Cerros, Belize; by David Pendergast at nearby Lamanai; and by Bruce Dahlin and Ray Matheny at El Mirador, Guatemala, have clarified much of the beginning of Maya power politics, for all show dramatically the elite use of pyramids as "political posters." The Late Preclassic structures at those sites and others were apparently adorned with huge stucco reliefs showing the visages of rulers in various mythical or cosmological guises. Evidently these impressive monuments were constructed to legitimize the rule of individuals and dynasties. Freidel notes a sudden change from an apparently egalitarian village of thatched dwellings to a larger town dominated by a stepped platform, adorned with these images of power. Perhaps, at this time, somebody got the idea of claiming descent from the sun and set about convincing his neighbors. At any rate, a real power elite was born, and its visible trappings spread quickly across the Petén and into Yucatán. We know neither the precise origins of all this, nor the nature of direction of its spread. Suffice it to say that there are incredibly massive structures at both El Mirador and its possible sister city, Calakmul, only about 19 miles (30 kilometers) away. The famed stucco-covered pyramid at Uaxactun, prosaically known in the literature as E VII-sub, is part of the same complex, as is the gigantic, almost perfectly preserved mask found a couple of years ago at the same site; the well-preserved masks at Kohunlich, Mexico; and the bas-relief of a ruler in full regalia at Loltun Cave, far to the north. The Maya, from the increasingly powerful elite dynasts to the talented farmers of maize, beans, and squash, were on the rise, so that, by say A.D. 250, the stage was set for the Classic Period florescence in every corner of the Maya area.

THE CLASSIC PERIOD (A.D. 250–900)

What date span should be assigned to the Classic Period is undergoing some modification, with a tendency to expand the period at both ends of the interval used here. That interval, based mainly on the use of Long-Count dating by the ancient Maya, doesn't really work well with respect to either the cultural developments in central Mexico; the rich cultural dynamics and accomplishments of so-called Preclassic peoples during the centuries centered on the beginning of the Christian Era; or the Classic-to-Postclassic cultural continuities evident in the archaeological record of the northern Maya Lowlands (the Puuc cities of Uxmal and others). In sum, it wouldn't be amiss to consider the "Classic" Period as extending all the way from around 300 B.C. to about A.D. 1200, and perhaps it will be so adjusted someday.

Teotihuacan

Now Mesoamerica's largest ruin, Teotihuacan was occupied from around 100 B.C. to A.D. 650, and toward the end of that span, it prospered as the largest city of the preindustrial world, with a population that may have exceeded 100,000—some would double that figure. Teotihuacan grew in the wake of the decline of Cuicuilco to the south, near the opposite end of the Basin of Mexico. The site of its founding appears to have been at least partly determined by the presence of the large cave now known to underlie the area where the dominant structure, the so-called Pyramid of the Sun, was erected.

Despite the size, obvious importance, and fame of Teotihuacan, relatively little is known about the inhabitants of the city with regard to ethnolinguistic identity. Part of this stems from the almost total lack of hieroglyphic writing at the site. What is clear, however, is that the place was the first great urban center of the Americas, and that its influence reached over much of Mesoamerica in the middle centuries of the Classic Period.

Teotihuacan. Avenue of the Dead with the Pyramid of the Sun in the background. By A.D. 200 this carefully planned and populous city was the most powerful political and cultural force in Mesoamerica. Photograph by Al Moldray, courtesy of the National Geographic Society.

Unfortunately, neither the nature nor the mechanism of this influence is known with certainty.

The mark of Teotihuacan in other regions of Mesoamerica is manifest in the distinctive greenish obsidian that was traded widely, and in various architectural and ceramic styles imitated in distant sites of the highlands and lowlands.

Teotihuacan itself was laid out on a grid and carefully planned so that its major axis lies almost exactly fifteen and a half degrees east of north. On this matrix, the "blocks" of the city held shops, craft workshops, and complicated, single-story "apartment" complexes, each with its living quarters, cooking areas, and religious shrines. Often the interior was decorated with mural paintings.

Excavations in certain areas of the great city have shown that there were "foreign" neighborhoods, where people from various areas of Mesoamerica lived and worked. Recent excavations in the heart of the Pyramid of Quetzalcoatl have revealed what appear to be burials of sacrificial victims. This and other research will, hopefully, begin to shed much-needed light on the elite power politics of Teotihuacan.

Teotihuacan declined abruptly sometime in the mid-seventh century, and the reverberations of its demise were felt throughout Mesoamerica, particularly in those areas of the ancient Zapotec, Maya, and others who had felt most strongly the influences of the powerful metropolis.

Monte Albán and Oaxaca

The great site of Monte Albán, carefully laid out on an artificially flattened mountaintop near present-day Oaxaca City, prospered as a Zapotec civic and religious center throughout the Classic Period. Its architecture resembles that of Teotihuacan, and overlies the earlier Preclassic remains at the site, mainly the Danzante reliefs.

The Zapotec remains at Monte Albán, as well as those of the nearby related sites of Lambityeco and Yagul, include low-relief carvings of figures and hieroglyphic texts and numerous elite tombs, often distinguished by their mural paintings and ceramic effigy incense burners.

Beginning around 500 B.C., but especially between about A.D. 300 and 750, Monte Albán gradually reached an estimated peak population of 25,000, dominating the other towns of the Oaxaca Valley. This major regional development is generally credited to the ancestors of the modern Zapotec, who still inhabit the area— and the great period of Monte Albán more or less defines the Classic Period of Oaxaca.

El Tajín and the Veracruz Area

The northern Veracruz cultures underwent their own version of Classic Period prosperity, mainly manifest in the huge site of El

Monte Albán. This monument of carved hieroglyphics is probably Zapotec. Photograph by George Stuart.

Tajín, notable for its distinctive architecture and apparent emphasis on the ritual ball game. The people behind the development and expression of Classic Veracruz culture might have been the ancestors of the modern Huaxtec or Totonac.

The Maya Highlands and the Kaminaljuyú Chiefdom

At some point during the early centuries of the Christian Era, the Kaminaljuyú chiefdom became connected with Teotihuacan, the distant city of Mexico's central highlands that flourished between A.D. 100 and 600, but neither the nature nor the intensity of the connection is known. The possibilities include military conquest by the Teotihuacaños; a trade arrangement; religious conversion; all of the above; or none of the above. Whatever it was, the archaeological record of highland and Pacific slope Guatemala suddenly reveals certain hallmarks of Teotihuacan, including the small, lidded cylinder tripod vases on slab legs, green obsidian from the Teotihuacan quarries, and distinctive architectural features.

Lowland Maya Civilization

The formal beginning of the Maya Classic Period is defined by the Long-Count date, equal to A.D. 292, on the back of Tikal Stela 29, the earliest known monument actually bearing a date which has been found in the Maya Lowlands.

At that time, we can envision a steadily growing population in virtually all corners of the Lowlands, with some sites, however, declining as the power politics and economics of the Late Pre-classic sort themselves out. The best example of this is the yet-unexplained eclipse of El Mirador, while nearby places of comparable size, such as Calakmul and Tikal, continue or begin to grow to enormous size. Whatever the details of the evolution of Maya civilization, its general nature during the period between around A.D. 300 and 900 continues to emerge from the published records of anthropological archaeology, epigraphy, art history, and other disciplines.

Politically, the Maya Lowlands—and most of our data come from the southern Lowlands—appear to have been divided into a patchwork of polities very similar in nature to the city-states of Greek antiquity. Each of these appear to have centered upon a sort of capital city; the larger states contained subsidiary cities and towns as well. As one might expect, the largest known of the ancient cities, such as Tikal, Cobá, and Calakmul, were such capitals. Others (among many) include Palenque, Piedras Negras, Yaxchilan, Bonampak, Uxmal, and Caracol. The boundaries of all these polities appear to have shifted through time as fortunes of their ruling dynasties rose and fell with the caprices of trade, agricultural prosperity, or warfare.

Much of Classic Period Maya warfare had a ritual aspect, and was apparently triggered by astronomical events such as the appearance of the planet Venus in certain positions. The inscriptions of the Classic Period are full of war and war-related events such as capture and sacrifice. The most prominent set of such texts are those of Dos Pilas and its close neighboring sites in the Petexbatun region of northwestern Guatemala, currently under investigation by Arthur A. Demarest and Stephen Houston of Vanderbilt University.

Of more than routine interest in the consideration of Maya warfare is a recently discovered altar from Caracol, Belize, for it notes the conquest of Tikal by Caracol.

Stemming from the great part of ritual or real war in elite Maya culture are such customs as the use in rulers' titles of a particular number of other elite personages they have captured (i.e., the epithet, "He of 20 captives" at Yaxchilan) and the widespread mention of specific episodes of capture.

Within a typical state, rulership lay in the hands of one or more elite lineages whose power was passed on in dynastic fashion. Often, such power and prestige were consolidated by marriage, even across political boundaries, as in the case of the "Woman of Tikal" who married a Naranjo ruler. Within their respective cities, members of an elite or ruling lineage dwelt in great complexes of masonry buildings, often painted or sculpted with the icons of the family and its real and mythical ancestors as posters of legitimate power. Over the generations—and if family power remained fairly stable—these acropolises grew as new construction was added. Old platforms and buildings were covered with new ones. Great pyramid-temples—architectural metaphors for sacred mountains and sacred caves—rose as memorials to dead leaders, and even these were covered up as more recent members of the lineage passed on to Xibalba, the Maya underworld. Away from the large palaces and public buildings, houses of pole and thatch occupied platforms of varying height, and in numbers that reflected the size of the extended family. Near these residential complexes lay gardens and household properties devoted to arboriculture, or the penning of deer, peccary, and wild fowl. On the outskirts of each

EMBLEM GLYPHS

The reconstruction of Classic Maya politics comes largely from the study of emblem glyphs in the hieroglyphic texts of the time. This written form, deciphered by Heinrich Berlin in 1958, now appears to refer to the states themselves and to a city only when that city is the only one in the polity, or simply bears the same name as the polity. Emblem glyphs (each of which combines several constant elements with a distinctive variable element) are known for most of the large cities of the southern Maya Lowlands, including Tikal, Naranjo, Palenque, Copán, Quiriguá, and Seibal. Variant emblem glyphs—those that lack the constants of normal emblem glyphs, but which nonetheless function in the same way, are known for Río Azul, Altar de Sacrificios, and other important Maya sites, thanks to work by Stephen Houston and Peter Mathews.

In important studies, Peter Mathews isolated a noble title that appears to designate the heads of subsidiary cities within a state, thus demonstrating a hierarchy of such places and, by extension, a ranking of their ruling families. Stephen Houston and David Stuart have isolated true place names for many cities in the northern and southern Lowlands, as well as names for areas of sites, individual buildings, and other features.

city lay the larger expanses of milpa (cornfield), sometimes on terraces, sometimes on the incredibly fertile rectangles of dried marsh deposits dredged from the grid of canals that defined the planting areas. In the canals, beneath the thick stands of waterlilies, lay the realm of the caiman and the turtle, two forms of life synonymous with the notions of earth and water—and thus fertility.

The teeming population of any Maya city was made up of a complex layered society with the minority elite at the very top.

Ultimate power in matters of war and ceremony lay with the ruler, whose legitimacy derived from the accumulated might of ancestors, both real and supernatural, whose names and times were carefully kept for recording on monuments and in buildings and in the hieroglyphic books. Members of the royal family, such as brothers of heirs to the throne or others who could not inherit the ultimate power, often became members of the prestigious classes of scribes, painters, architects, and sculptors. Other members of Classic Maya society included merchants, potters, masons, jewelers, and the all-important farmers. The talent of the latter was of particular relevance, given the precarious agricultural potential of most of the Maya area.

However impressive the expression of Classic Maya civilization at places like Palenque, Río Azul, or Copán, there seems to have been an inherent flaw in the system, for the whole began to come apart at the seams sometime in the ninth century. Perhaps the elite games of ritual conquest and sacrifice had gotten out of hand; maybe the elite were finally revealed as pompous martinets and the lower strata of society rose up against them; perhaps the agricultural capacity of the land was irrevocably reduced by the clearing of the forests or by climatic change. Most likely, the "collapse" of the Classic Maya resulted from a complicated series of events, a chain-reaction involving crops, trade, power shifts, perhaps even invasion, and the ever-precarious state of the general economy so typical of a land where ideas outnumber natural resources—and where one can succeed but once in trading an idea. Or, the so-called collapse may be, at least to some degree, a false archaeological problem, more a construct of the modern investigator than a cataclysm of Maya reality. Thus, a cessation of construction of large masonry buildings and the abandonment of the custom of erecting dated monuments could indicate problems confined to the elite and not to the general populace. Thus, many Mayanists advocate a balance in the targets of research, which could be accomplished by more emphasis on the digging of non-elite structures.

At the present time a team of investigators from Vanderbilt University, under the direction of Arthur Demarest, is in the midst

of a five-year program centered on Dos Pilas and nearby Guatema-
lan sites in the area of Lake Petexbatun. The purpose of the project
is to determine the possible role of warfare in the collapse. To
date—and all their material is still under analysis—the team has
documented evidence of sieges and hastily built defensive walls at
Dos Pilas, along with an entire new site, Punta Chimino, which
had a defensive moat and palisades. Demarest and his colleagues
have also found new inscriptions dealing with the history of warfare and conquest between Dos Pilas and its neighbors, including
distant Tikal.

Whatever caused the Maya collapse, its effects are most apparent in the southern Lowlands and the adjacent Usumacinta Valley.
In those areas, the evidence of the elite activity that gave primary
definition to the Classic version of Maya culture—the erection of
dated stone monuments, the building of great stone funerary
edifices and other public buildings, and the enlargement of the
sprawling architectural complexes that define the core areas of the
great cities—appear to have ceased within a century or so. In the
north, the Classic cities (with the exception of Cobá) appear to
have moved without much ado into the Postclassic Period.

THE POSTCLASSIC PERIOD
(A.D. 900–1521)

Central Mexico

After the fall of Teotihuacan and the evaporation of its power by
A.D. 700 or so, Mexico's central highlands became the setting for
numerous centers and polities whose individual power was more
localized. Among them, the great city-center of Cholula, in the
adjacent Valley of Puebla, a minor counterpart to Teotihuacan
during the Classic Period, prospered. And others, from Azcatpotzalco and Texcoco, on or near the great system of lakes south-

west of the abandoned Teotihuacan, to nearby Tlaxcala and distant Tula, prospered as well.

The main story of central Mexico during the Postclassic Period revolves around two great cultures—the Toltec, centered on their capital at Tula, who, according to later Mexican Indian oral tradition, dominated the area all the way to the Maya Highlands from about A.D. 800 to 1000; and the Mexica, known to posterity as the Aztec, who migrated to the Basin of Mexico in the twelfth century and established themselves on Lake Texcoco. In time, these latecomers gained the advantage of power politics, and founded their capital, Tenochtitlán, on an island in the lake, in 1325.

From Tenochtitlán, the Aztec welded a tribute empire that stretched from the Basin of Mexico to the very edge of the Maya area. It was this empire and its capital that fell to the soldiers of Cortés during the momentous conquests of 1519 to 1521.

Oaxaca

Meanwhile, Postclassic Oaxaca became the stage of interaction between the Mixtec, who occupied the mountains west of Monte Albán, and the Zapotec, who dominated a zone centered on the Oaxaca Valley. What appears in the archaeological record to be, in effect, a general Mixtec incursion into the Oaxaca Valley in the eighth century marks the beginning of the Postclassic Period.

At Monte Albán burials of the Mixtec elite now fill many of the old tombs emptied of earlier Zapotec nobility. Most famous among these is Tomb 7, discovered by Mexican archaeologist Alfonso Caso in 1932—a Mixtec burial chamber filled with remarkable objects of gold, semiprecious stone, and other treasures, which constitutes perhaps the greatest single archaeological treasure unearthed to date in all of Mesoamerica.

The metal ornaments from Monte Albán Tomb 7 and other Mixtec burials in the area bear witness to extraordinary talent in metalworking, one of the traits that distinguishes between the Classic and Postclassic periods, for metal and knowledge of its working appear to have been introduced into Mesoamerica around A.D. 1000.

One of the great Postclassic centers of Mesoamerica is manifest in the elaborate architectural complex at the ruins of Mitla, the traditional Mixtec hub throughout most of the period.

Veracruz

The Postclassic history of the Veracruz area is little known. Suffice it to say that eventually the fall of the great Classic Period center at El Tajín gave way to other, apparently smaller polities that became incorporated into the Aztec Empire in the century of so before the coming of the Spaniards.

The Lowland Maya

While Tikal and other large cities of the southern Lowlands appear to have been abandoned by around A.D. 900, others continued—at least in ways that are apparent in the archaeological record. Topoxte, on Lake Yaxha, seems to have thrived during the Postclassic Period, as did Lamanai, and a handful of other centers in the region. But even in enduring whatever the general social upheaval was that lay behind the collapse, these places reflect fundamental changes.

If the Classic Maya were ruled by an elite whose prolific commissions of buildings and monuments dominate the visible record, the Postclassic Maya differed enough to be readily apparent in the remains themselves. True, the great buildings are still there, as witness the so-called Palace of the Governor at Uxmal, or the Castillo and Great Ball Court at Chichén Itzá. But what is added to Maya culture itself results from a shift in emphasis: More is happening in the realm of carefully managed trade, in military activity, and in the general Mexicanization, or at least internationalization of the Maya world. This character pervades both Lowland and Highland communities to such a degree that, in perusing

The Castillo, or Temple of Kukulkan, at Chichén Itzá is one of the largest pyramids in the Yucatán Peninsula. Photograph by Barbara Lerner Kopel.

Intricate geometrical relief work characterizes the Puuc style of architecture at Uxmal. Photograph by Barbara Lerner Kopel.

the earliest Spanish accounts of the Conquest among, say, the Highland Maya capitals of Utatlán and Zaculeu, it is difficult to distinguish the descriptions of architecture, costume, and other superficial features from those of the Aztec and others in central Mexico. This Mexican problem has vexed archaeologists for decades.

Anyone can see that the Puuc architecture typical of Late Classic–Early Postclassic Uxmal is different from that of, say, Palenque or Tikal, but regional styles are common in the history of Maya architecture.

The Yucatec Maya word Puuc meaning "hill" or "range of hills," names the low ridge that diagonally crosses the area just north of the western segment of the boundary between Yucatán and Campeche states. It also gives its name to a style of ancient Maya architecture characterized by complex stone-mosaic facades dominated by large masks with long curving noses, generally agreed to be representations of the Maya rain god Chac (also rendered as Chaac, Chak, or Chaak in Yucatec). Puuc architecture characterizes Uxmal, Kabah, Labna, Sayil, and numerous other ruins in the area immediately south of the ridge.

Two other architectural styles should be mentioned here as well. The Chenes style, named for the region of natural wells (*ch'en* in Yucatec means *well*) south of the Puuc region, is marked by central doorways made to represent great monster mouths, probably representing an earth god as a cave. The Río Bec style of Maya architecture, named for a site where it was first noted, features two or three solid towers symmetrically arranged with regard to the structure they are part of, but constructed as false, nonfunctional pyramid-temples.

Some archaeologists and architectural historians believe that these styles have chronological as well as geographical significance, and postulate a Chenes-Puuc sequence around A.D. 700 to 900.

Some see central Mexican influence in the complex mosaic facades of the Puuc region—influence further suggested by the presence of skulls and bones on the bas-reliefs of platforms. Others are wary of the connection. It is at Chichén Itzá, however,

Detail from the Monjas Quadrangle at Uxmal depicting the Maya rain god Chac. Photograph by Otis Imboden, courtesy of the National Geographic Society.

The Caracol, or Observatory, at Chichén Itzá. Photograph by Barbara Lerner Kopel.

the famed site that lies far to the east of Uxmal, that the Maya–Mexican question comes to full flower. In terms of architecture, part of that site (old Chichén) is virtually indistinguishable from Uxmal and other Puuc cities, while the other part is visibly different—featuring feathered-rattlesnake columns, long colonnades, and other distinctly non-Maya traits. This set of differences, plus the Colonial Period chronicles that tell of what may have been an invasion by Mexicans, plus the close resemblance of non-Maya Chichén Itzá to distant Tula, the traditional Toltec capital in Mexico's central highlands, has fostered the current view by most of a Toltec invasion of Yucatán around the end of the tenth century. Perhaps, but, as was the case at Kaminaljuyú, we simply do not know enough yet to say what actually did happen at Chichén Itzá in the turbulent years before its abandonment, which tradition sets at A.D. 1200. To the contrary, work by Charles Lincoln strongly suggests that the fundamental ethnic makeup of Postclassic Chichén

Itzá did not change, but remained Maya throughout the span that embraces the incursion of the Itzá—probably other Maya who settled the place around A.D. 900 (and gave it their name)—and the putative invasion by the Toltec a century or so later. This evidence lies mainly in the consistency of motifs in the art, despite some apparent change in style, and the fact that new phonetic decipherments of the hieroglyphic texts of Chichén Itzá reveal Yucatec as the elite language, even during the so-called Toltec Period.

Our general view of the northern Lowlands between, say, A.D. 900 and 1200 reflects the old city-state plan—a series of provinces with primary and secondary centers. Among the capitals of this Early Postclassic Period, we have Uxmal, Izamal, and many others on the Yucatán plain from Kan Pech (now Campeche) to Kutzumil (Cozumel). And while the boundaries of the provinces doubtless shifted as the fortunes of their respective cities rose and fell, they probably coincided generally with those which are known for later times. Ralph Roys (1957) offers the best treatment of this subject of the late polities of prehistory in the northern Maya Lowlands. Some modifications of his scheme have been indicated by subsequent archaeological research: The work of Anthony Andrews of New College, in Sarasota, Florida, at the site of Isla Cerritos shows that tiny island at the very northeast tip of the peninsula was probably the seaport for Chichén Itzá. If so, the provincial boundaries were indeed different early on, for the boundaries documented just after the Conquest would not have permitted that capital a "window to the sea."

In general terms, the Early Postclassic Period in the northern Lowlands reflects an emphasis on trade and militarism. The first is shown by the distribution of both goods and art styles; the latter by the masses of heavily armed soldiers who adorn the columns and facades of Chichén Itzá.

The final centuries of Maya hegemony that preceded the appearance of the Europeans saw the rise of Mayapan, a densely populated walled city north of declining Uxmal, and west of Chichén Itzá, which, if we can believe the legends, was already returning to the forest—except for the activity of pilgrims who still went there on special occasions to toss jewelry, incense, or their fellow believers into the Sacred Well.

Mayapan must have been an amazing place. Its central precinct, the seat of those powers that controlled the combination of politics, religion, and economy of much of western Yucatán, was dominated by unrefined but fairly sturdy imitations of Chichén Itzá's Mexican-style Castillo and colonnades. And within the great wall that encompassed the city lay more than 4,000 structures, most of which were dwellings that, had they survived, would have been indistinguishable from the thatch-and-pole houses we see today in nearby Telchaquillo and the other northern Maya towns. The high-status masonry residences of Mayapan were airy places, most often fronted by columns and centered by darker vaulted chambers for sleeping and other domestic activity. Some are decorated with Puuc-style rain-god masks, and all were heavily stuccoed and often painted in deep reds, light blues, yellows, and other colors that held both sacred meaning and aesthetic appeal.

As a regional power, Mayapan endured for a brief two centuries or so, until about 1450, when the later chronicles note its demise, apparently aided by a disastrous drought. Most of the other great stone cities of the north shared in the erratic history of the Postclassic. Uxmal and its neighbors in the Puuc region, although largely abandoned in the tenth century, maintained a small-scale population, sometimes bolstered by newcomers like the Xiu family who, in Colonial times, would claim Uxmal as their ancestral home. Grass and vegetation completed their takeover of the proud edifices of Chichén Itzá while the former inhabitants of the city—according to tradition and the chronicles—established and consolidated a new city, Tayasal, far to the south, on the lake that now bears their name—Petén Itzá (Island of the Itzá).

Meanwhile, on the eastern coast of Yucatán—roughly the area now embraced by Quintana Roo state—a chain of towns and small cities prospered as the great trade canoes moved cloth, salt, pottery, and other goods in and out of ports of call that punctuated the area from Cozumel southward to Nito and other centers in present-day Guatemala and Honduras. Christopher Columbus sighted just such a trade canoe on his fourth voyage (1502). Peter Martyr, in

The Temple of the Frescoes at Tulum is decorated on two of its four corners with the rain god Chac. Photographs by Barbara Lerner Kopel.

his important *De Orbe Novo* (various editions, 1516 on) describes the cargo Columbus saw as including "bells, razors, knives, and hatchets made of yellow or translucent stone; there were also household utensils for the kitchen, and pottery of artistic shapes . . . ; and chiefly draperies and different articles of spun cotton and brilliant colors" (MacNutt translation, 1912). This episode also marks the first use of the term *Maya* (as "Maia") in the literature of Europe—apparently in reference to an area of present-day Honduras. In the manuscript of Bartholomew Columbus, which describes the same incident, the word is *Maiam* with the words *vel luncatan* (Yucatan?) superscribed in a different hand.

Among these latest of the Postclassic cities of Yucatán are Tulum, Tancah, and the large city of the same date that sprawled over the old Classic Period ruins of Cobá. All exhibit buildings in the distinctive east coast architectural style—relatively rude masonry structures with small (and sometimes illogically tiny to the modern observer) doorways in which globs of stucco helped make up the loss of regularity in columns, walls, and facades. Many structures bore carved stucco ornamentation and murals, both interior and exterior, depicting supernatural beings, people, birds, and animals.

This is not to say that mural-painting was more common to the Postclassic than the Classic Period, only that the later examples are more frequently preserved. One factor in this is, of course, the fact that, where they occur, Postclassic structures represent the last phase of building before the Conquest. Among the best examples of Postclassic Maya painting are those of Tulum (particularly the Temple of the Frescoes), Cobá, and Xelha. The largest single set of murals, that which adorned the exterior of the intentionally buried Postclassic building at Santa Rita Corozal, Belize, discovered by Thomas Gann in the late 1800s, is now gone, preserved only in the discoverer's copies published by the Smithsonian in 1900.

As might be expected, the Postclassic Maya paintings reflect, in the manner of their depictions of figures and adornments, influences of central Mexico, particularly the Mixtec style known from Oaxaca at this time. One can, in fact, speculate that free-lance muralists from the west were among those traveling the trails and canoe routes of Postclassic Mesoamerica. Examples of Classic Period murals of any size are quite rare. The notable exception lies in the amazing paintings of Bonampak, Chiapas. There, three entire rooms were covered from floor to capstone with brilliant paintings showing a late eighth-century ceremony—probably connected to the presentation of the heir-designate of the ruler Chaan-Muan, whose reign over the area ended with the demise of elite activity at the site in the 790s. The paintings, discovered in 1946, are duplicated in various places (the whole set at Mexico's National Museum of Anthropology, and Room 1 at the Florida State Museum in Gainesville) but none of the versions adequately duplicate the detail or richness of the originals.

For archaeologists, one of the hallmarks of this final period of the pre-Hispanic Maya is the presence of large, brightly painted ceramic effigy censors whose broken sherds cover the surface of the last Maya occupations of the old cities from Mayapan to Cozumel, and from Campeche into the Petén. These often portrayed gods in the form of standing humans, holding the accouterments and wearing the insignias and garb that identify them, and backed by cylindrical receptacles. Set in household shrines or ranged upon altars fronting the painted temples, these sacred vessels held the holy pom (incense) that played a crucial role in Maya (and Mesoamerican) ritual life from the very beginning. In the year 1517, at a Maya coastal town whose name is forever lost—explorer accounts call the place Gran Cairo—such effigies were filling the air with oily resinous smoke, and staring blankly at the turquoise waters of the Caribbean when Spanish ships of the Córdoba expedition appeared in the vicinity of Cape Catoche in the spring of 1517. Its landfall, either at or near the cape itself or at Isla Mujeres, just to the north of present-day Cancún, is in dispute, but authorities generally favor the former. As for the effigy incense burners, so numerous were they that the Spaniards, who believed them to be small representations of females, gave the place its present name, meaning Island of Women.

THE COLONIAL PERIOD
(A.D. 1521-1821)

For both Maya and Spaniard, the Conquest was a long affair. Exploration of the coast of the Yucatán Peninsula and the Gulf Coast of Tabasco were accomplished by Grijalva in 1518. A year later Cortés's expeditionary force paused briefly at Cozumel before the great journey that culminated in the conquest of Tenochtitlán, the capital of the Aztec, on the site of present-day Mexico City, in 1521—the year that marks the official boundary between the Postclassic and the Colonial periods in Mesoamerica.

In 1524, the dreaded Pedro de Alvarado laid waste to the Highland Maya capitals of Guatemala, including Utatlán and Zaculeu. The conquest of Yucatán by Francisco de Montejo culminated in the founding of the city of Mérida (atop the Maya city of Tiho) in 1542. Final consolidation of the Spanish presence was not accomplished for another century and a half, when the Itzá city of Tayasal, deep in the Petén, came into the Spanish fold in 1697. As a consequence of this complicated chain of episodes, the interior of the Maya Lowlands, particularly its southern reaches, remained terra incognita for many decades.

Cortés visited Tayasal on his epic overland journey from Mexico City–Tenochtitlán to the trade city of Nito (on Lake Izabal near the coast of Guatemala) in 1525–1526. The famed Fifth Letter of the conquistador describes in detail the trip, which did not result in any lasting conquests of the interior.

In 1988, four sheets of a late seventh-century manuscript were discovered in a private library in Mexico. Ultraviolet photography revealed it as a fragment, in Spanish, of a hitherto unknown trip by Spaniards and Maya Indians from Tipu to Tayasal around 1695, two years before its conquest. This important document is now in the library of Mexico's National Institute of Anthropology and History. The document contains the only known eyewitness account of a Maya ruler (Can Ek of Tayasal) in full pre-Conquest regalia.

SHIPWRECK SURVIVORS MEET CORTÉS

A well-known encounter took place on Cozumel Island between the Cortés party and the Spaniard Gerónimo de Aguilar who, with his companion Gonzálo Guerrero, had survived a 1511 shipwreck on Alacranes Reef. Taken captive, both survived, Aguilar as a slave, while Guerrero eventually became a respected war leader. The latter, according to the account by Bernal Diaz del Castillo, was married, had three sons, and had, in effect, become a Maya. After the brief meeting with Cortés, Aguilar vanished into history. The fate of Guerrero is unknown.

Except for recent work at Tipu, Belize, by Robert Kautz, Grant Jones, and Elizabeth Graham, the archaeological picture of the Colonial Period is scanty. As a consequence, our knowledge of culture and the process of the blending of Maya and European ways relies almost exclusively on the documents of history. The most important of all of these, indeed, the primary source for the study of Mexican culture, is the remarkable *Relación de las cosas de Nueva España,* compiled by Fray Bernardino de Sahagún in the mid-1500s. It provides us with a detailed view of Aztec culture at the time of the Conquest and affords many insights into Mesoamerican culture in general. Its counterpart, for Maya culture, is the account by Bishop Diego de Landa, who wrote what has come down to us as the *Relación de las cosas de Yucatán,* an extraordinary document compiled around 1566. The original manuscript by Diego de Landa is lost. The work, probably used in its original form by the historian Herrera in his great history published between 1601 and 1615, survives as an abstract in the archives of the Academia Réal de la Historia in Madrid. That precious manuscript remained in obscurity until 1863, when the indefatigable French cleric Brasseur de Bourbourg found it. It was published in Paris the next year, but incompletely. All dozen or so subsequent editions are flawed in one way or another, but the best (in English) is that under the editorship of Alfred M. Tozzer, published by the Peabody Museum of Archaeology and Ethnology at Harvard in 1941. The work is laden with nuggets of information about the Maya, including an illustrated guide to the hieroglyphs which, through work by the Russian scholar Knorozov in the 1950s and others since, has allowed the phonetic decipherments by today's working epigraphists.

Landa's Yucatán was a world of Maya small-town maize farmers only slightly overlain by the growing influence of Spain. The settlements of pole-and-thatch houses were often cornered by stone cairns that duplicated the boundaries of the cosmos. Rites and rituals paced the successive stations of the native calendar, drawing upon a myriad of divinities to witness the sacrifices and the divinations guided by tables in hieroglyphic books which, along with the "idolatrous" incense burners and bizarre statuary, the Franciscans sought to eliminate. Despite the obvious differences between the two worlds that confronted one another in sixteenth-century Yucatán, certain underlying similarities actually aided the visible adjustment: The old Maya city arrangement—obvious in the mysterious ruins concealed in the forest—adapted to the grid-and-plaza standards of Spanish town-planning. The numerous gods matched rather conveniently with the saints of the Catholic church. And the pomp and ceremony of the mass must have seemed, to the Maya fresh from the ritual of the cornfield, a familiar thing indeed.

Beneath the Mesoamerican facade, however, much went on. In the Maya area and elsewhere, heart sacrifice, the ultimate effrontery to the Christians (despite the zealous barbarities of the Conquest itself), persisted in isolated areas well into Colonial times, and most of the motivation underlying the performances of native rites has survived to the present. So have the Indian languages— Nahuatl, Otomi, Mixtec, Zapotec, and many others. The primary languages of the Maya Lowlands in pre-Conquest times were apparently Yucatec and Cholan, with the boundaries shifting through time. The dictionaries and vocabularies compiled by various

Spanish cleric-scholars in the early centuries of the Conquest are of immeasurable help in the ongoing work directed at the decipherment of ancient Maya writing. Though changed some through time, much as English has evolved since the time of Chaucer, some thirty versions of Mayan are still spoken. In 1821, exactly three centuries after the Conquest of Mexico, Spain lost its American dominions to the revolutions of national independence inspired by Simón Bolivar.

THE MODERN MAYA WORLD

The Modern Period has seen the continuation of much of Maya life in its traditional form, but subjected to the pressures of national and state politics that have, at times, exceeded the bounds of public tolerance. An eighteenth-century uprising in Chiapas was short-lived, but the later Maya rebellion in Yucatán was not. The War of the Castes, as it is known to historians of the period, began in 1847 with Maya raids against Valladolid and other towns of the northeastern peninsula. For more than half a century, the region seethed with sporadic engagements, gruesome massacres, and the religious fanaticism of the Cruzob, the followers of the Talking Cross. Mexican national politics, completely in disarray over state and federal rivalries—these were the days of Santa Anna (b. 1795?; d. 1876), Emperor Maximilian (b. 1832; d. 1867), and the aftermath of the war over Texas—proved ineffective in stemming the revolution. As a consequence, the Maya actually came close to recapturing the entire northern part of their old area. The war, however, petered out toward the end of the century. In the end, General Bravo and his army reached the Maya capital, Chan Santa Cruz (The Little Holy Cross)—now the town of Felipe Carrillo Puerto, Quintana Roo—and ended the bloody era.

One of the great unsung sagas of history, the war was best chronicled by Nelson Reed in *The Caste War in Yucatán*. As a result of this strife, the whole eastern portion of the Yucatán Peninsula—the territory which, in 1975, became the state of Quintana

This roadside cemetery in Quintana Roo illustrates the convergence of cultures in the Yucatán Peninsula. Photograph by Barbara Lerner Kopel.

Roo—was effectively depopulated by the early years of this century except for small isolated Maya towns. Until the 1940s outsiders generally feared to enter the zone. Ironically, perhaps, the modern resort of Cancún, and the ruins of Tulum and Cobá, have now attracted a hitherto unimaginable number of outsiders to this area, virtually none of whom know of its grim history.

Throughout this century the Maya world and that of the *dzulob* (the non-Maya) have continued to coexist in varying degree, depending upon whether the setting is rural or urban. In Chiapas, the Highland Maya of the area around Zinacantan (like many of the Pueblo groups of our own Southwest) continue to protect their privacy well, though they have been extensively reported upon by anthropologists who have devoted much time to their languages, settlement patterns, belief systems, and remarkable textile arts. The Lacandon Maya, who dwell in the forest that bears their name, slowly increase in number, but face the simultaneous obliteration of both their culture and their forest as logging and resettlement of other Maya penetrate their realm. And the northern Maya of Yucatán, like all the others of recent times, slowly adjust to the arrival of electricity, new roads, and—if great ruins lie nearby—tourism. Some four million Maya-speakers today inhabit the area of their ancestry, although their demographic and ethnolinguistic distribution differs in many important respects from those of earlier times. The Petén, where it all may have begun for the Maya, is just now being reoccupied in greater numbers as homesteaders—many from the Highlands—appear in lands vacant for a millennium. The flow of refugees, strong during recent years of unrest and oppression, has subsided and even reversed in some cases, but the result nonetheless is a veritable crazy-quilt of polyglot settlements in the former frontier areas of Mexico and Guatemala. In such places— the wilderness of the Lacandon Forest, the dry scrub country of central Campeche and lower Quintana Roo, and others—the pristine vegetation, whether jungle or savannah, burns to create space to accommodate ranches, cornfields, and living space. And with the flames perish the macaws, the howler monkeys, the sacred jaguars, the deer, and the peccary. And with the fire, the power of which seems to rival even the awesome countenance of Lord Sun, also die dozens, perhaps hundreds, of ancient ruins yet unnamed or unseen by outsiders, for the heat transforms their limestone monuments and walls into powder. Thus do the writings on stone—intended to immortalize some of the greatest rulers of American antiquity—vanish, unread, as civilization of quite another sort encroaches upon the lands of the Maya and their neighbors.

PART TWO

The Itinerary

Mexico City

Lagunilla
Market

Templo
Mayor

Bellas Artes   Cathedral

Ave. Madero

Torre
America
Latina

Zócalo

Ave. Juarez

Monument to
the Revolution

Ave. Insurgentes

Paseo de la Reforma

Zona
Rosa

Melchor Ocampo

Paseo de la Reforma

National Museum
of Anthropology

Museum of
Modern Art

Chapultepec
Castle

Teotihuacan

México

Vaso del Lago de Texcoco
(circa A.D. 1519)

México
Districto
Federales

Volcán Itzaccíhuatl
(1605 m)

Volcán Popocatépetl
(1682 m)

Morelos

Volcán Ajusco
(1199 m)

Volcán Chichinautzin
(1059 m)

△ *Day One*

MEXICO CITY

As you will notice from any flight approaching Mexico City—that sprawling metropolis of 25 million, ever in contention for the status of "largest in the world"—the city occupies much of the floor of a huge natural basin surrounded by volcanoes and volcanic debris. Its elevation is 7,349 feet (2,240 meters).

At the time of Hernán Cortés's arrival in 1519, this Basin of Mexico (more popularly known as the Valley of Mexico) was centered on a system of large lakes. And on an island in Lake Texcoco, one of the largest lakes, lay the Aztec capital Tenochtitlán, and its sister city Tlatelolco. Causeways connected the island city with various other places on the shore. Almost nothing remains of the lake systems. Indeed, much of the land now occupied by Mexico City was reclaimed from the lake system during Colonial times, and some of the old Aztec causeways became principal avenues.

The international airport lies on the eastern fringe of the city. After your landing approach, you will descend through a yellowish haze and taxi into the international arrival gates. The Mexico City airport is an east–west linear affair subdivided into sections A through D. As you proceed along the crowded promenade, you will see numerous "hole-in-the-wall" shops on your right, interrupted by the doors to the taxis, limos, and traffic outside. Among the shops are two of unusual merit: the Cartografía store in the area of section B, where your can get great maps of Mexico and its regions, and the Antropología shop, near the end, in section A, where you can get official site guidebooks and other publications of Mexico's Instituto Nacional de Antropología e Historia (INAH) on Mexican archaeology, along with excellent and inexpensive replicas of ancient and Colonial art. Their stock varies from visit to visit. Some of the same material, and more, is also for sale at many museums and at the regional offices of INAH in Oaxaca, Yucatán, and other states.

Just outside the international arrivals section is a window where taxi tickets are sold. You can pay your fare in advance and your ticket will be honored by any of the cabbies who hover around the terminal. A tip is not expected. If you do not get a ticket, be sure to agree upon the fare with the driver before you get into the taxi. Bus is the most popular means of transportation in Mexico City, but also the most crowded. The metro is an excellent system and goes to the airport, but you are not allowed to bring suitcases and large bundles into the system. (The rule is not enforced before 7:00 a.m. and after 9:00 p.m.)

Mexico City has been laid out in a north–south, east–west grid pattern since the days of the Aztecs. The area the city now occupies was in Colonial days occupied by numerous villages. The villages are now incorporated into the city as *colonias,* or neighborhoods, but each has retained its own names for its streets. Now there are about 350 *colonias* and many of the streets in Mexico City have several different names as the street passes through the various neighborhoods; be sure you know the district in addition to the street address before you set out.

The Zócalo, the traditional center of the Spanish community, is officially called the Plaza de Constitución. Around it are the National Cathedral, City Hall, the Capitol, and the President's office. Monte de Piedad, the national pawnshop (with branches all over Mexico) is a more recent addition, as is the Museo del Templo Mayor.

About eight blocks west of the Zócalo is Alameda Park—which began life as an open space for such spectacles as the burning of heretics during the Inquisition. From the northwest corner of Alameda Park runs Paseo de la Reforma, Mexico City's widest street. It passes southwest through one of the city's most affluent business and tourism districts—the *zona rosa* or pink zone—until it intersects Chapultepec (literally grasshopper hill) Park, which contains seven musems, lakes, an amusement park, and an old battlefield. Infinite options greet the visitor to Mexico, even one with rather specialized interests. Numerous archaeological sites occur in the immediate area of Mexico City and a short distance away in the Basin of Mexico. Here we will treat only the major places that afford more than routine insight into the archaeology of the country—the Museo Nacional de Antropología (National Museum of Anthropology), the new Museo del Templo Mayor (Museum of the Great Temple), and the famed archaeological site of Teotihuacan.

National Museum of Anthropology, one day

The National Museum of Anthropology, on Paseo de la Reforma in Chapultepec Park, is one of the world's greatest museums, in terms of both architecture and the richness of its content. Designed by architect Pedro Ramírez Vásquez, the magnificent building was completed in 1967. There are five basic parts to the museum: entry/foyer, exhibit halls, outside exhibits, ethnography halls, and visitors' facilities. It is open Tuesday through Saturday, 9:00 a.m. to 7:00 p.m.; and Sunday 10:00 a.m. to 6:00 p.m.; closed Monday. There is a small admission fee; free on Sunday.

The large, open entry/foyer area holds the excellent museum shop (to the left as your enter); the admission-ticket counter (to the right); and special exhibits, including the "piece of the month" (also to the right). The shop sells books, museum reproductions, postcards, and other items that you might expect. Here, however, the selection is more complete than usual. Packages and other such impedimenta may be checked nearby. Special permits must be obtained for photography, and the use of flash is not permitted. Access to the offices of the director and staff of the museum, and the library and other facilities, is restricted to varying degrees, and may be reached from the foyer area after a proper check-in with the guards on duty.

CUICUILCO

The important Preclassic site of Cuicuilco lies in the southern reaches of Mexico City, 1.8 miles (3 kilometers) beyond the National Autonomous University of Mexico, set in the Parque Ecológico Cuicuilco just off Insurgentes to the left and immediately after the intersection with the Perférico Sur. The site is dominated by a circular, three-step "pyramid," which marks an important Middle-to-Late Preclassic site now covered by several meters of lava flow from the volcano Xitle, which erupted around 200 B.C.

Santa Cecilia, a restored Aztec building, and the great Chichimec pyramid (and site) of Tenayuca, nearby, mark the northern zone of the city.

The site is open Tuesday through Saturday from 10:00 a.m. to 5:00 p.m. and Sundays and holidays from 10:00 a.m. to 4:00 p.m. There's a small admission fee to the museum, but it's free on Sundays and holidays.

The twenty-one *salas* or exhibit halls surround the gigantic patio that lies beyond the entry area. This magnificent open area is dominated by a massive column-fountain that alone supports a huge square roof that partially covers the patio. Each of the exhibit halls is devoted to a particular ancient culture area of Mexico. The great Sala Mexica (at the far end) holds many of the famous remains of the Aztecs, including the famed Calendar Stone, or Sun Stone.

Outside exhibits lie adjacent to some of the first-floor exhibit halls. These include architectural reproductions as well as full-scale replicas of monumental sculpture. Other sculptures are set up here and there on the edges of the patio.

Upstairs, the second-floor ethnography halls contain the rich panorama of modern Mexican Indian life and culture.

On a lower level, accessible from the south side of the great patio, are the visitors' facilities, including restrooms and a restaurant, which is open daily from 9:00 a.m. to 7:00 p.m.

Museo del Templo Mayor, three hours

The relatively new Museum of the Great Temple of the Aztecs lies in the center of Mexico City at the Zócalo, beside the actual site of the complex of structures, including the Great Temple itself. These buildings formed the heart of Tenochtitlán, the Aztec capital between A.D. 1325 and the Spanish Conquest of 1521.

The vast excavations beside the museum, between it and the Cathedral, hold the foundations of many of the Aztec temples and other constructions that were built here. Of these, the lower portion of the Great Temple itself is both the most imposing and the most complicated, for excavations showed that the huge double-pyramid was made of numerous earlier stages of construction. All these remain for the visitor to see. Access to the stairway remnants, painted altars, and sculptures is afforded by a system of catwalks that reach virtually every corner of the great excavation.

The Templo Mayor museum, a magnificent example of modern architecture, also holds models of the ancient capital and many examples of sculpture found in the excavations of the site between 1978 and 1982. It is open Tuesday through Sunday, 9:00 a.m. to 5:00 p.m. and closed Monday; admission is free.

Teotihuacan

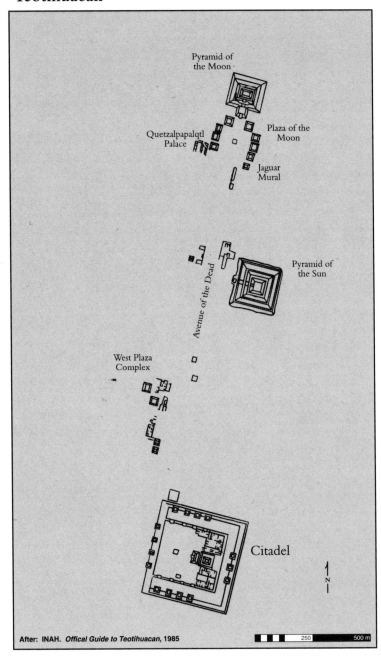

Pyramid of
the Moon

Plaza of the
Moon

Quetzalpapalqtl
Palace

Jaguar
Mural

Pyramid of
the Sun

Avenue of the Dead

West Plaza
Complex

Citadel

N

After: INAH. *Offical Guide to Teotihuacan*, 1985

250   500 m

△ *Day Two*

TEOTIHUACAN

Teotihuacan is the largest archaeological site in all of Mexico, and one whose known past had an enormous effect on the cultural history of greater Mesoamerica. It is about 30 miles (50 kilometers) outside Mexico City to the north via Avenida Insurgentes Norte to the Pachuca toll road (Route 85) (and the famed Colonial church of Acolman and the small exhibit on the Paleoindian remains of Tepexpan Man).

To take the bus from Mexico City to Teotihuacan, buy a ticket at the Terminal de Autobuses del Norte. This bus station is at Avenida de los Cien Metros, 4907, which is about a third of a mile (half a kilometer) north of Insurgentes Norte, and 3 miles (5 kilometers) north of the Zócalo. A metro station is just outside the door. The ticket counter you want is at the far left as you enter, and buses leave every twenty minutes from 6:00 a.m. to 6:00 p.m. The trip takes about an hour, and the bus drops you at the gate of the site. You catch the return bus where you got off and pay the driver's assistant.

In the Mexico City area, the great site is known simply as the Pyramids. The builders of Teotihuacan laid the great city out on a grid whose central axis, known as the Avenue of the Dead, was laid out on a bearing of fifteen and a half degrees east of north. The Avenue of the Dead begins at the plaza fronting the Pyramid of the Moon and proceeds toward the southwest for about 1.2 miles (2 kilometers), past stepped platforms, compounds, and palace

Teotihuacan. The Pyramid of the Moon is faced with tiered platforms linked by a 150-foot staircase. From its end at the Avenue of the Dead, the pyramid presides over more than 100 shrines and temples. This, America's first metropolis, held sway over much of Mexico from A.D. 100 to 700. Courtesy of the National Geographic Society.

complexes that line both sides, and eventually vanishes into the landscape of a portion of the site that remains largely unexcavated.

The bulk of the ruined city, away from the Avenue of the Dead, consists of city blocks, defined by the ancient grid pattern of streets. Excavation reveals apartment-like complexes holding living quarters, kitchens, patios, shrines, and remnants of splendid mural paintings. Tepantitla, where part of one of these suburban residential complexes has been restored, features the famed painting of Tlalocan, the earthly paradise of the rain god. Other such complexes open for visit include Zacuala, Tetitla, and Yayahuala.

The easiest way to see "downtown" Teotihuacan is simply to follow the Avenue of the Dead from the Pyramid of the Moon to the great enclosure known as the Citadel, centered by the ornate Temple of Quetzalcoatl. In between lie the Palace of the Plumed Butterfly, the huge Pyramid of the Sun, and other structures, all of which repay a visit.

The complex known as the Palace of the Plumed Butterfly, excavated and paritally restored in the early 1960s, features pristine bas-reliefs, some with inlays of obsidian, used as architectural decoration. Farther along, the Jaguar's Palace holds beautifully preserved paintings on its intact lower walls.

The Temple of the Sun, largest structure at the site, was erected over a cave system, the latter discovered only within the past twenty years. The pyramid itself, perhaps the largest in all of Mesoamerica, was originally somewhat larger. Unfortunately, its early "restorers," lacking the techniques of today, unwittingly removed most of the last layer of facing stones, so the overall aspect of the whole is not quite as it was originally. One occasionally reads that the Pyramid of the Sun is larger than the Great Pyramid of Giza, in Egypt. Not so. Although the basal measurements of the two are nearly the same (some 750 feet, or 230 meters, on a side), the height of the Pyramid of the Sun (originally about 245 feet, or 75 meters) is but half that of the Egyptian structure.

The Temple of Quetzalcoatl, which is decorated with richly detailed sculptures of serpent heads and rain-god countenances, has recently been investigated by archaeologists from Mexico's National Institute of Anthropology and History. In the depths of its central core, they unearthed the remains of what may have been a large elite tomb, which was looted in antiquity. Surrounding it, along the four sides of the pyramid base were the skeletal remains of what appear to be bound soldiers, perhaps captives sacrificed as retainers.

Oaxaca Valley

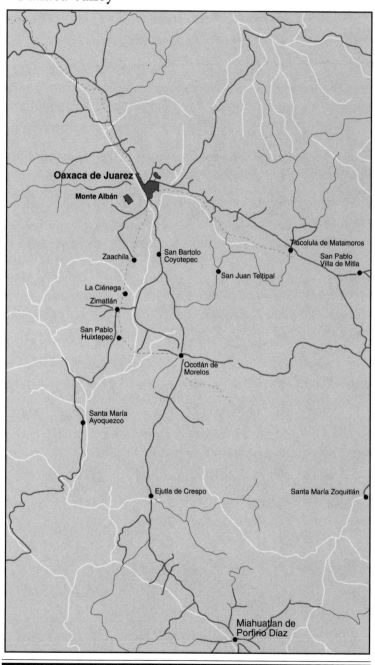

△ Day Three

OAXACA

Mexico City is 320 miles (510 kilometers) from Oaxaca City (population, 150,000) in Oaxaca state. The bus takes nine hours. The train leaves Mexico City at 5:32 p.m. and arrives in Oaxaca at 8:05 a.m. The flight is much shorter and affords a splendid view of the incredible topography of Mexico's central highlands, for it traverses the outermost portion of the Mexican Plateau and the jumbled highland-and-valley systems to the south of the plateau edge. The Oaxaca Valley is one of these.

Oaxaca is warm and dry most of the year. Half of the annual twenty-four inches (sixty centimeters) of rain falls between June and September. The area is the main producer of *mezcal,* maguey cactus liquor bottled with a maguey worm.

Of the 2.7 million people who inhabit the state, 1 million (a fifth) are Indians who belong to at least seventeen different groups. Two thirds of the immediate area's population is either Mixtec or Zapotec. As a consequence, Oaxaca is notable for its rich traditions of the textile and ceramic arts. The famed Saturday Indian markets are the largest in Mexico.

Oaxaca, built by the Spanish and originally named Antequera (it became Oaxaca in 1532), retains much of its Colonial charm in the cobbled streets and the centuries-old architecture of the cathedral (constructed in 1544), churches, and homes. In spite of earthquakes (notably in 1560 and 1854) and Indian rebellions, Oaxaca had a population of 2,000 by 1620, and by the late 1700s, it was probably the third largest city in New Spain. By this time, it had grown rich by exporting cochineal, a red dye made from insects. But central to this labor-intensive manufacturing and trade was the virtual enslavement of some 30,000 peasants who produced the dye for traders who lent them money. In 1783, when the Spanish crown banned this abuse, the cochineal boom went with it.

Cortés himself loved the area and claimed much of the valley for himself, along with the title Marques del Valle, but his planned estate was never built. The area also provided Mexico with two of its greatest political leaders—Benito Juárez (b. 1806; d. 1872), a Zapotec Indian who became president of Mexico in 1859, and Porfirio Díaz (b. 1830; d. 1915), who ruled the country between 1877 and the Revolution of 1910.

Oaxaca City boasts several notable museums. Of particular interest is Museo Regional de Oaxaca (the Regional Museum), housed in a former convent adjoining the church of Santo Domingo, about seven blocks north of the Zócalo. This museum holds the remarkable Postclassic Mixtec treasure excavated from Tomb 7 at Monte Albán—among the greatest archaeological finds ever recovered in the Americas.

△ Day Four

OAXACA VALLEY SITES

In very general terms, the Oaxaca Valley is shaped like an inverted Y, with the city of Oaxaca (elevation, 5,084 feet, or 1,550 meters) and the ruins of Monte Albán at its central juncture. The Etla Valley, running northwest–southeast, forms the main stem of the Y. The Tlacolula arm, east of Oaxaca City, holds the archaeological sites of Mitla, Lambityeco, Dainzú, and many others.

Monte Albán, a half day

The ruins of Monte Albán (White Mountain) lie on a flattened mountaintop some 1,200 feet (400 meters) above the valley floor and around 5.5 miles (9 kilometers) outside Oaxaca City. The site may be reached by a slow and scenic route that winds up the slope. From the top, the view of the valley is unparalleled.

To get to the site, you can take a taxi from the Zócalo, or take a bus (run by Autobus Turistico) from the Hotel Mesón del Angel on Mina between Díaz and Mier y Terán.

Along the road to Monte Albán, you may see numerous local vendors offering "antiquities" for sale. A few are genuine; most are not. Whatever the case, avoid the issue by NOT purchasing any of this material, for taking antiquities out of Mexico is both wrong and illegal.

The pyramids, temples, and ball court of Monte Albán were carefully laid out on a north–south plan, and placed along the edge

Monte Alban

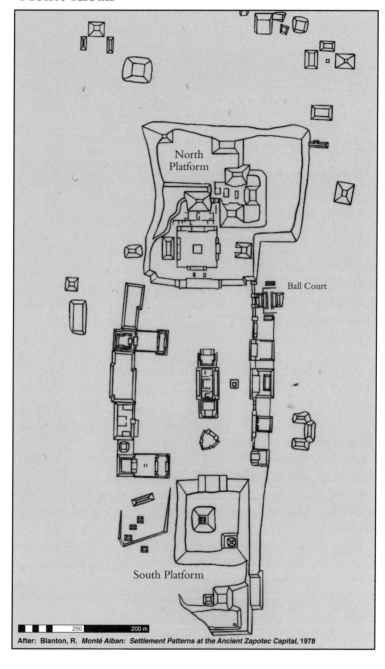

North Platform

Ball Court

South Platform

After: Blanton, R. *Monté Alban: Settlement Patterns at the Ancient Zapotec Capital,* 1978

250   200 m

Monte Albán. The plaza—nearly the size of four football fields—is flanked by stepped pyramids that conceal tombs. Monte Albán, the ceremonial center of the Zapotec, rivaled Teotihuacan. From here priests and princes ruled from A.D. 100 to 900. Courtesy of the National Geographic Society.

of their mountaintop setting so as to surround a vast cental plaza. Architecturally, the structures—which generally date to the Classic Period—recall those of Teotihuacan. In addition, numerous remains date from the earliest known occupation of the site (around 600 B.C.). These are visible mainly in the form of the so-called danzante (dancer) carvings—old structural stones bearing low-relief sculpture that appears to have been reused in later construction.

According to the best current interpretations, the danzante stones represent not dancers, but dead, mutilated, sacrificial victims—probably chiefs of villages conquered by the Preclassic rulers of Monte Albán. Some of the depictions are accompanied by hieroglyphics, in all probability the names or places that pertain to the persons in question. Other reliefs of the same time hold short glyphic texts that include dates, and are the earliest texts of this kind in Mesoamerica.

Of the two major structures that occupy the central part of the plaza, Mound J is of special interest, for it is unique at the site in both orientation and plan. The structure is pentagonal and aligned on an axis about forty-five degrees east of the main site axis, with its "point" to the southwest. For this reason, it is said to have functioned as an observatory, though the contention is unproven. The walls of the Mound J structure hold many slabs, reused, bearing columns of Zapotec hieroglyphs.

Trails lead from the main plaza of Monte Albán to other ruins on the upper slopes of the mountain. Among these are the remarkable Tomb 7, where the great treasure was excavated in 1932, and Tomb 1905, which holds well-preserved Classic Period murals.

Other Sites

Other archaeological sites in the Oaxaca Valley that are of more than routine interest include Dainzú, Yagul, Lambityeco, Zaachila, and Mitla.

MITLA, TWO TO THREE HOURS

The important site of Mitla (from the Nahuatl *mictlan,* meaning place of the dead) is about 28 miles (45 kilometers) from the Zócalo. You can reach it by taxi or by bus. The bus leaves from Trujano station. The ruins are open from 8:30 a.m. to 6:00 p.m. The small admission fee is waived on Sunday.

The site is mainly Mixtec, featuring extraordinarily well-preserved buildings with intricate mosaic facades in complex geometrical patterns. Here, one sees the true genius of the ancient Mesoamerican architect.

YAGUL, LAMBITYECO, AND DAINZÚ

Yagul, Lambityeco, and Dainzú can be reached by car by continuing past Mitla, northwest on Route 190.

Turn right off Route 190, 6.2 miles (10 kilometers) past Mitla, and continue for 1.2 miles (2 kilometers) for Yagul. Return to the main road, continue 3 miles (5 kilometers), turn right, and continue 2.5 miles (4 kilometers) to the Lambityeco site. Yagul and Lambityeco (the former in a lovely hilltop setting), near one another in the eastern arm of the Oaxaca Valley, show Zapotec temple and tomb architecture of the Classic Period.

For Dainzú, return to the main road and continue past the Lambityeco turnoff 4.4 miles (7 kilometers), turn left and drive 0.6 mile (1 kilometer) on an unpaved road to the site. At Dainzú, Preclassic bas-reliefs (similar in style to the Monte Albán danzantes) depict ball players in full costume, with protective masks and gloves, and holding balls about the size of modern baseballs.

Zaachila, about 11 miles (18 kilometers) south of Oaxaca City, was the last Zapotec capital, and features interesting tomb bas-reliefs.

Villahermosa to Palenque

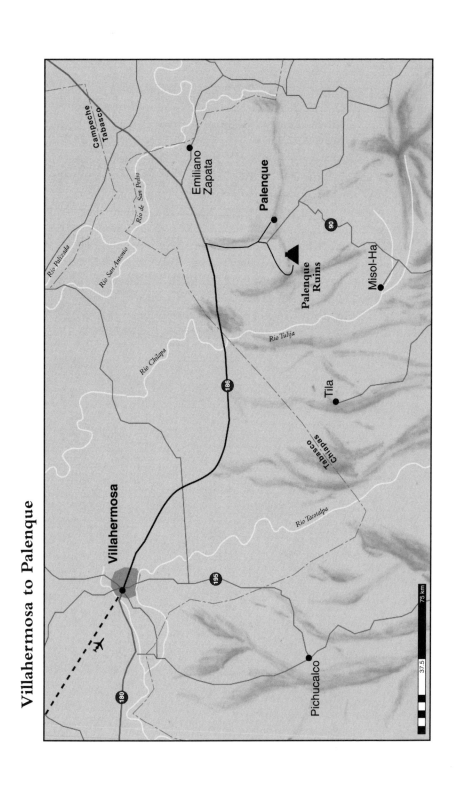

△ *Day Five*

PALENQUE

The flight from Oaxaca to Mexico City retraces the route taken earlier, and connection to Villahermosa is made from there. The flight between Mexico City and Villahermosa often provides a spectacular view of the volcanoes Popocatépetl (Smoking Mountain) and Iztaccihual (Sleeping Lady) and, farther on, the peaks of Malinche and snow-clad Orizaba, Mexico's highest mountain. The destination, Villahermosa, capital of the state of Tabasco, lies on the hot, humid Gulf coastal plain, beside the Grijalva River. The trip to Chiapas takes three and a half hours and can be made by A.D.O. bus from Villahermosa (an early morning bus is necessary to avoid trouble in getting a hotel room) or by cab. Cab fare is generally inexpensive; however, be sure to negotiate the fare before entering the cab. The journey takes us generally west on the main Mexico City–Mérida highway; enter the state of Chiapas, and turn south at kilometer 113 toward Palenque. At about 3.7 miles (6 kilometers) and on the left lies a tiny unexcavated set of mounds that, because of the nature of the archaeology of this region, could be anything from Preclassic Olmec to Late Classic Maya. From here the highway winds over a rolling green countryside largely devoted to ranching.

Soon the horizon begins to reveal the jagged gray silhouette that marks the northernmost face of the Chiapas highlands—a set of parallel limestone ridges covered with rain forest punctuated by encroaching clearings for farms and pastures. The ruins of Palenque lie on a kind of shelf area on this face of the mountains and are

Palenque

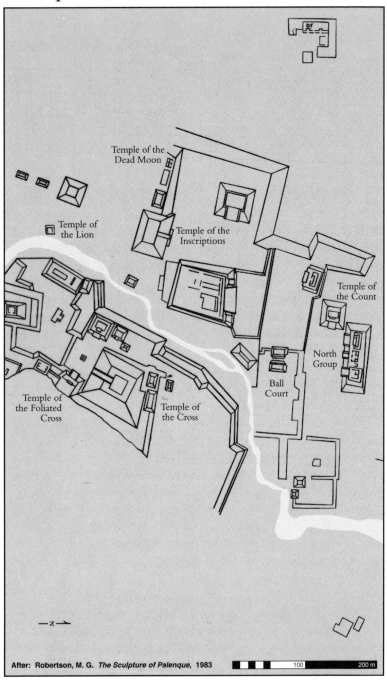

Temple of the Dead Moon

Temple of the Lion

Temple of the Inscriptions

Temple of the Count

North Group

Temple of the Foliated Cross

Temple of the Cross

Ball Court

—z→

After: Robertson, M. G. *The Sculpture of Palenque*, 1983

100   200 m

sometimes visible from the highway just before the huge U-curve across the Mexico City–Mérida railroad tracks at Palenque station. Several miles farther, past the Palenque airstrip, we come to the fork featuring the gigantic sculpture of a Maya head appearing to rise from the ground. The left fork of the road leads through the modern town of Palenque (population 35,000), to the Hotel Misión Palenque on the south fringe of the settlement; the right fork leads to the famous ruins.

The town was founded in the 1570s, but the now-famous ruins were not discovered until the mid-1700s, when the dense forest on the slope to the west yielded mysterious "stone houses." The ruins were examined several times in the 1780s by small expeditions sent from Guatemala City–Chiapas was then part of Guatemala— and Baron Alexander von Humboldt (b. 1769; d. 1859) published one of the stucco carvings in 1810, but attributed it to a ruin in Oaxaca state. In 1822, Palenque became the subject of the first illustrated report on a Maya ruin, and has been in the mainstream of research ever since, from the time of the eccentric Constantine Rafinesque (b. 1783; d. 1840) (who published on the hieroglyphics in the 1820s and 1830s) to the era of the late Mexican archaeologist Alberto Ruz who, in 1952, made the greatest discovery of all at the site.

The Ruins, three and a half hours

Many consider Palenque the most beautiful of all Classic Maya sites, with Uxmal and Copán as close seconds, but each place is different and you must decide this matter for yourself. However it ranks in this respect, the place thrived from the fourth to the eighth centuries, reaching its height of power and extent under the rule of Pacal the Great (reigned A.D. 615–683).

Palenque has several unique traits. For one, the ancient architects solved the problem of combining a large roof comb, or flying facade, with large rooms, themselves limited by the confining nature of the Maya false arch, or corbelled vault. First, they slanted the upper facades back between the eaves and roof, lessening the

Palenque. In the distance to the left is the Temple of the Inscriptions, and to the right, a part of the North Group. Photograph by Otis Imboden, courtesy of the National Geographic Society.

masonry load on the walls below. Second, they moved the roof combs to the center axis of the building so that the great weight pushed directly down on the center wall. Third, they built tall vaults with niches. The grand result is manifest in buildings with the large doorways—the largest in the Maya area—and rooms of great width, both serving to create the illusion of light openness despite the mass of stone involved. Another mark of Palenque is the lack of stelae, the upright slabs commonly used throughout Classic cities to record the images and texts of royal power. Instead they used great limestone panels set into the interior walls of the buildings, perhaps taking advantage of the very fine-grained limestone that occurs locally. Still another quality of Palenque derives from the use of carved stucco to decorate many of the wall surfaces. These have been miraculously preserved, despite 1,200 or so rainy seasons that have pummeled the buildings.

Palenque. The Temple of the Inscriptions at the end of the path is characterized by a broad stairway and three-story tower. The Palace complex is on the left. Palenque, in the highlands of Chiapas, flourished under the rule of Pacal (see frontispiece) and his descendants. Courtesy of the National Geographic Society.

After negotiating the twisting road up the mountain, we park and enter the ruins along the gravelly path that passes Alberto Ruz's tomb to the open area fronting the Temple of the Inscriptions and the west facade of the Palace. The Temple of the Inscriptions contains the Pacal tomb. If you visit the tomb, remember that the descent is long and tiring and the stone stairs are extremely slick with moisture.

The main portion of our Palenque time should perhaps be taken up by simply wandering the vaulted passageways and patios of the Palace itself—a huge acropolis with a bewildering series of superimposed structures, including the unique Tower. By the base of the Tower is House E, evidently the coronation room for Pacal and others of the Palenque royal line. It is marked by the Oval Tablet set into the rear wall, showing Pacal receiving the headdress of rulership from his mother, Lady Resplendent Quetzal.

Southeast of the Palace, across the clear stream of the Otolum and the ancient aqueduct, lies the Cross Group, including the Temple of the Sun, probably the best-preserved building at the site. The Cross Group as a whole commemorates the ruler Chan-Bahlum (Snake-Jaguar), son of Pacal the Great, and traces him from accession to power to death and entry in Xibalba, the Maya underworld.

The museum, down the path past the Palace and Ball Court, contains some of the finest panels and stucco carvings found over the years of excavation. These include the Palace Tablet and the famous Tablet of the Ninety-six Glyphs, a sort of Palenque king list found between the Tower (where it was set into the foundation) and House E. The museum is poorly lit, so carry a flashlight if you want to see details of these exhibits.

Palenque offers much more than the ruins themselves. The forest would take volumes to treat with justice, and it shelters areas of stunning beauty, from the rocky stream of the Otolum to the magnificent stepped waterfalls near the highway leading up the mountain. Another fall and pool lie in the forest behind the museum. The forest also shrouds an area of ruins whose extent exceeds the open area that forms the usual limit for tourist visits. Here, too, are the trails that lead south into the hills and beyond to the towns of the Chol and Lacandon Maya. You may see the latter, dressed in traditional whitish sack-dress, selling their bows and arrows near the entrance to the ruins.

At La Cañada in December 1973, the first Palenque Round Table was held. Attended by 25 scholars, it resulted in the first of the breakthroughs in Maya hieroglyphic writing that have marked the years since. That meeting, and others that take place every two or three years, were inspired by Merle Greene Robertson, who lived at La Cañada from 1970 to 1983 with her late husband Bob. Their house, "Na Chan-Bahlum," is opposite the restaurant, and the street is officially named for her.

The La Cañada motel complex also witnessed the revival of stone carving at Palenque, mainly by the huge Morales family that holds three or so distinctive hotels and restaurants here. Using the same fine-grained limestone that the ancient Palencaños employed in their sculptured panels, the modern carvers make small-scale versions of the most famous reliefs from Palenque and other sites in the Usumacinta drainage, and offer them for sale. The work is mostly of high quality—so much so that the Reader's Digest book *Mysteries of the Past* illustrated the sculpture outside Carlos Morales's restaurant (near La Selva) as an ancient work! These carvings are well worth close inspection.

Return to the Villahermosa airport.

Mérida

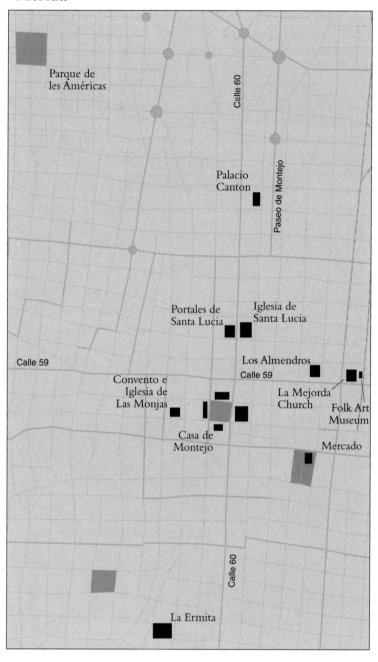

Parque de les Américas

Calle 60

Palacio Canton

Paseo de Montejo

Portales de Santa Lucia

Iglesia de Santa Lucia

Calle 59

Los Almendros

Calle 59

Convento e Iglesia de Las Monjas

La Mejorda Church

Folk Art Museum

Casa de Montejo

Mercado

Calle 60

La Ermita

△ *Day Six*

MÉRIDA

The flight from Villahermosa to Mérida crosses the southeastern part of the Bay of Campeche, then crosses the shoreline of the northwest portion of the Yucatán Peninsula. The aerial vista of the landscape here shows the flat limestone plain that typifies this northern portion of the Maya Lowlands. And the closer to Mérida, the more obvious are the hallmarks of this distinctive region—the network of *albarradas* (dry stone walls) that subdivide the surface of the land. Some enclose vast fields of sisal (henequen), the "green gold" of Yucatán, that provided cordage for the world, and prosperity to the area, until the advent of synthetics a generation or so ago. Other walls define house lots, farmsteads, and ranches, often marked by the distinctive round-ended thatched houses typical of the zone.

The Mérida airport lies just beyond the southeastern fringe of the city, near the industrial complexes and cement plants. The main route to the center of town is the Avenue of the Itzas, which sweeps north to El Parque Centenario, the local zoo and amusement park, and beyond. To arrive at the center of town, turn right at Calle 59 and proceed due east to the area centered on 60th Street. This is the central zone of the Noble and Loyal City of Mérida, founded in 1542 by the Conquistador Francisco de Montejo, the Younger, and built with the stone of the Maya city of Tiho, which originally occupied the spot.

The first thing to know about Mérida (population, 600,000) is that nearly all its streets are numbered. Odd numbers run east–

Horse-drawn carriages are a common sight in the lovely Colonial city of Mérida. Photograph by Barbara Lerner Kopel.

west, even numbers, north–south, so it's virtually impossible to get lost. Calle 60 is the main north–south axis, and Calle 59 is probably the most useful east–west line for a mental reference grid. The Zócalo (Plaza Mayor) is a point of reference, bounded by Calles 60, 62, 61, and 63. Most of the first-class hotels in the city lie near that reference point or along the Paseo de Montejo, a huge avenue punctuated by monuments, statuary, and old mansions that runs north–south (parallel to, and between, Calles 56 and 58).

The Palacio Cantón, the turn-of-the-century mansion that houses the Yucatán Regional Museum of Anthropology, is on the Montejo, and has splendid exhibits of archaeological material that will supplement what we have seen so far and enhance that which is to come.

Mérida's best shops are spread along the Paseo de Montejo, and clustered in the area of several blocks around the intersection of

DZIBILCHALTÚN

About 13 miles (22 kilometers) north of Mérida on Highway 261, just east of the Progreso Highway, lie the ruins of Dzibilchaltún, investigated and mapped by the National Geographic Society–Tulane University project of 1957–1962 under the direction of E. Wyllys Andrews IV. The site is unremarkable save for its great expanse, the intact Temple of the Seven Dolls, a pretty cenote, a restored Colonial Period *capilla abierta* (open chapel), and a nice small museum.

Calles 60 and 59. Of particular note are the "fashion versions" of Yucatec Maya *huipiles* and the men's *guayabera,* a variant of its Filipino counterpart, and the male clothing of choice for virtually any occasion.

By far the best regional food in the area—though competition is growing—is served at the restaurant Los Almendros, on the east side of town near one of the great Colonial Period gateways to the city. Opposite it, on Calle 50 between 57 and 59, lies the church of La Mejorada, and behind that, east a few yards on Calle 59 and virtually hidden from view and general knowledge, is an excellent museum of Mexican folk art, with a sales shop.

△ *Day Seven*

UXMAL

The bus for Uxmal leaves the city of Mérida via the Avenue of the Itzas and passes the international airport en route to Uman and points south. Just before reaching the airport, we pass a small, inconspicuous outdoor exhibit on the right, at the offices of the local highway department, showing a reconstruction of the cross-section of one of the ancient causeways of Cobá. On it has been placed the original of one of the huge limestone "rollers" found, still in place, on the Cobá–Yaxuná causeway some years ago.

At Uman, notable for its massive church, the road forks. The main branch is the relatively new and direct highway to Campeche; the left is the old road to Uxmal.

Yaxcopoil, an hour

The tiny settlement of Yaxcopoil (Place of the Green Poplar Tree) is interesting for two reasons. It is centered on one of the more prominent old henequen haciendas of the area, and there is a large archaeological site nearby. In the vast henequen fields just east of the settlement—and you may be able to glimpse the huge mound from the road—lies a large and relatively unexplored archaeological site of the same name. It has some standing architecture and at one time murals were visible. These were destroyed in the 1950s. Aerial photographs and satellite imagery reveal the site of Yaxcopoil as having concentric walls and a *sacbe* that leads from one part of the

Henequen fluorishes on the thin soil of the Yucatán Peninsula. It transformed the Yucatán economy of the 1880s, enabling plantation owners to live in luxury. Courtesy of the National Geographic Society.

site to another. The place, like hundreds of others, awaits more thorough examination.

Yaxcopoil lies within the main henequen zone of Yucatán, and the local people here, who harvest and process the leaves from the fields on the Yaxcopoil lands, have turned the old estate into a museum—one of the few such things available for the visitor to the area. One may tour the big house and wander the large layout of outbuildings and processing works. This crop once made the people of Yucatán—at least those who owned the haciendas—rich. In the days when the spiky henequen, or sisal, plant provided much of the material for the world's rope, the plants seemed to stretch to infinity. Each field, bounded by unbelievable lines of dry-stone wall and carefully mapped, had its own name for the record, and was subdivided into a vast sixty-five-foot (twenty-meter) grid marked by cairns. Here the rows of plants were nurtured and tended, and at intervals the bottom rows of long spiky leaves were harvested and sent to local processing plants on platforms that traveled a network of narrow-gauge rails. At such processing plants, special machinery separated the tough fiber from the pulp; the world bought it eagerly; and Yucatán prospered.

The golden age of henequen lasted less than a hundred years and peaked in the decades around the turn of the century, when the high society of Mérida and the surrounding haciendas—the families Peón, Molina, and others whose names still fill the Mérida phone book—looked to France for fashion, furniture, and the education of their children. Meanwhile their Maya workers cut and baled the precious fiber, ran clanking machinery that might serve as a local model of the Industrial Revolution, and spread the useless pulp in ever-expanding areas where it lay and filled the air with its sweet stink of decay. Each hacienda had its main house with glistening floors of polychrome tile, high-beamed ceilings, and recessed hammock hooks of silver. Each also issued its own money—a sort of company store coinage in such variety that the hacienda tokens of Yucatán formed the topic of the first memoir of the American Numismatic Association.

That did not last. Revolution and, ultimately, the introduction of synthetics struck deadly blows to the system and the market. Today, as you will see, henequen is still grown, cut, and processed, under the auspices of the Cordemex agency. But more and more of the fields are yielding to weeds or to alternative experiments with citrus orchards and anything else that might grow on this waterless landscape where limestone bedrock lies less than half-a-plowshare-blade from the surface.

Uxmal Ruins, three hours

The small town of Muna lies at the foot of the Puuc, the prominent ridge that cuts across the peninsula from northwest to southeast, dividing it between the flat limestone plain of the extreme north, where we have been, and the slightly undulating country to the

Uxmal

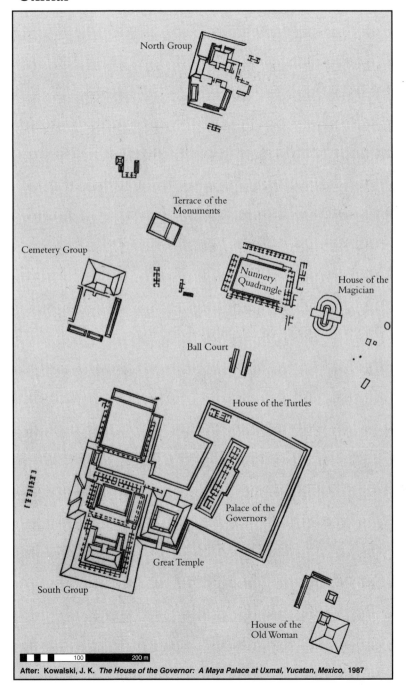

North Group

Terrace of the Monuments

Cemetery Group

Nunnery Quadrangle

House of the Magician

Ball Court

House of the Turtles

Palace of the Governors

Great Temple

South Group

House of the Old Woman

100   200 m

After: Kowalski, J. K. *The House of the Governor: A Maya Palace at Uxmal, Yucatan, Mexico,* 1987

View from the Monjas Quadrangle at Uxmal. The Ball Court, Palace of the Governor, Great Pyramid, and South Temple are all visible. Photograph by Barbara Lerner Kopel.

The Monjas Quadrangle (Nunnery) at Uxmal. Photograph by Barbara Lerner Kopel.

One of the most provocative structures at Uxmal is the elliptical Pyramid of the Magician. Photograph by Barbara Lerner Kopel.

A side view of the Pyramid of the Magician at Uxmal (right). A portion of the Monjas Quadrangle is on the left. Photograph by Barbara Lerner Kopel.

south, where Uxmal and the other famous Puuc sites lie. The raised terrace beneath the church at Muna is ancient and contains potsherds that date mainly from the Classic Period. After winding our way briefly over the ridges of the Puuc, we descend and proceed south to Uxmal, 6.2 miles (10 kilometers) distant. Our first view of the ruin comes at the crest of a hill, framed in the front windshield—a set of massive gray buildings silhouetted against the blue sky. Here, at the sharp turn that takes the highway farther into the Puuc region and on to the Campeche (and eventually to Mexico City), lie what some consider the most beautiful ancient buildings in the entire Maya area.

Uxmal, another site whose ancient name is known from the chronicles and histories, was brought to prominence by the popular travel account published by John Lloyd Stephens (b. 1805; d. 1852) in 1843, and by the exquisite and accurate drawings by Frederick Catherwood, who illustrated that work. The property was then part of the vast chain of haciendas owned by the Peón family of Mérida, and Stephens and Catherwood stayed at one whose ruins still lie in the mosquito-plagued bush behind the present Hacienda Uxmal luxury hotel (just inside the curve made by the highway). There are now several other hotels near the archaeological zone, and a new visitors center–museum, as well, for the region is uncommonly rich in interesting and well-preserved ruins of the Late Classic and Early Postclassic Periods, among them Kabah, Labna, Xlabpak, and Sayil. All are famed among architectural historians as examples of the Puuc style, which features incredibly complex mosaic facades and beautiful veneer masonry.

As is the case with most sites in the Maya area, Uxmal is much bigger than the cleared and manicured (to varying degrees) area open to visitors. And at least one sacbe is known, leading to Kabah, about 12.4 miles (20 kilometers) by highway. Our visit will generally be confined to that part of the site where there is the most to see, namely, the Pyramid of the Magician, the Nunnery Quadrangle, the Ball Court, and the massive platform holding the House of the Turtles and the Palace of the Governor. These are names of convenience, partly founded in local myth of unknown validity.

Uxmal itself is the seat of the Xiu family, famous in the early Colonial Period, whose descendants still live in the region. A look at these structures will prove infinitely more rewarding than any words we can muster to describe them. Among the things to note: The representations of Maya thatched houses incorporated into the south range of the Nunnery; the incredibly steep and complicated superimposition of structures and styles that makes up the House of the Magician; the elegant simplicity of the House of the Turtles; and the sheer mass and design of the Palace of the Governor. The latter is the largest ancient building (not counting platform) in Mesoamerica, and it was set so as to face directly certain heliacal risings of the planet Venus, whose symbols appear in its intricate facade.

Return to Mérida

The return to Mérida will take us either back the way we came or by means of an alternate, longer route via Kabah, Loltun Cave (a gigantic cavern open to tourists, where mammoth bones have been found), and the historic towns of Maní (where Bishop Landa lived) and Chumayel (where the most famous Maya chronicle was found). Also along the route lie the ruins of Mayapan (which, after Uxmal, may be anticlimactic), and the town of Acancéh (which boasts many ruins within its limits, including an Early Classic pyramid facing the main plaza).

Mérida to Uxmal, to Chichén Itzá

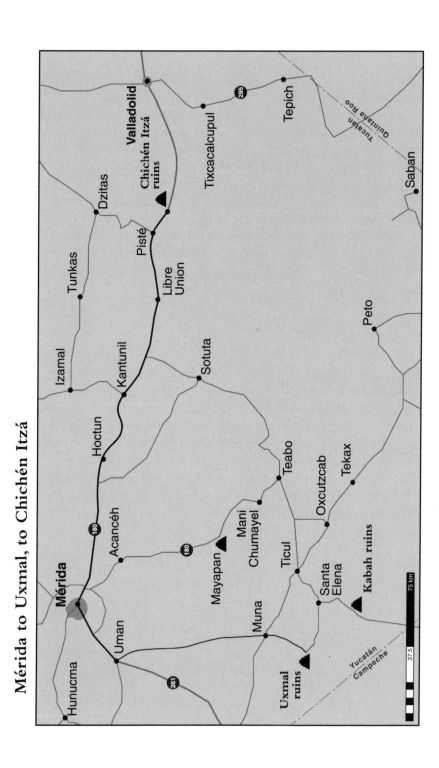

◬ *Day Eight*

CHICHÉN ITZÁ

From Mérida, the trip eastward toward Chichén Itzá (75 miles or 121 km on Highway 180) takes the traveler through several towns of slightly more than routine interest, including Libre Union and Pisté. All such settlements, by the way, will have good examples of traditional Maya houses—oval in plan with a single central door and tall thatched roof. Walls may be of small vertical poles tied together, and may be coated with mud or lime plaster and white-washed. You may also be able to glimpse sleeping hammocks hung in the interior, either functioning or tied against one wall for the day. Often a second structure, for cooking, lies just behind the main house. Yards are enclosed by the ubiquitous dry-stone walls, made of limestone boulders cleared from nearby fields. Every yard will have its laundry center, marked by a *batea* (washboard)—a long horizontal basin balanced between two supports. Until fairly recent times these were of wood, but due to the depletion of the trees—you will notice that there are few really big ones—the law now dictates that these be concrete.

You will also note the dress of the northern Lowland Maya. Women wear the traditional *huipil*, a white beltless sack-dress with decorated neck and hem. These vary in elaborateness from simple store-bought versions to intricate hand-embroidered or cross-stitched models. The most elaborate—for ceremonies and other special occasions—are short, in order to reveal an intricately worked underskirt, and set off by gold jewelry, for Yucatán is noted for its filigree work. Men mostly wear blue jeans and denim

Chichén Itzá

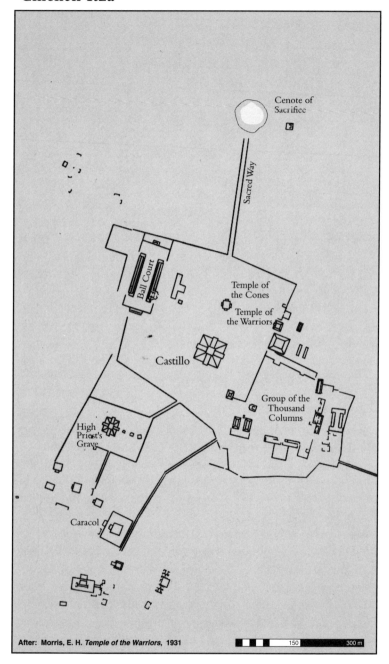

Cenote of Sacrifice

Sacred Way

Ball Court

Temple of the Cones

Temple of the Warriors

Castillo

Group of the Thousand Columns

High Priest's Grave

Caracol

After: Morris, E. H. *Temple of the Warriors*, 1931

150   300 m

shirts. The older men often retain the short white pants and shirt of earlier decades, sometimes punctuated by a panel of pin-striped cloth in front. All wear straw hats. Men often carry guns (for hunting), a woven shoulder bag of henequen (for the day's food), and a hollow double gourd with stopper (for water).

Izamal and Kantunil

At Kantunil, we turn left and drive to Izamal (population, 20,000). Our visit there places us almost simultaneously in the ancient past of the Maya Classic Period, the Colonial splendor of sixteenth-century Yucatán, and a reasonably prosperous modern town. The marks of all three appear quite clearly to one flying over Izamal, for its street plan is interrupted by a massive Early Classic terrace topped by one of the largest pyramids in the northern Lowlands, and by the great expanse of the atrium of the Franciscan convent built between 1553 and 1561. It is the latter that mainly concerns us here.

The place is a minor architectural wonder, particularly considering the setting and the time it was erected, with local labor in newly conquered Yucatán. Of note is the interior of the church proper, and the stone passageways that lead through the complex. In a small room on the top story at the rear hangs one of only two known portraits of Diego de Landa, third bishop of Yucatán and compiler of the famed *relación* that provides anthropologists with the best single account of the northern Maya at the time of the Spanish Conquest. From the top of the staircase leading to the room, you can enjoy a splendid view of the nearby pyramid. The large cross on top is the focal point of local ceremonies honoring the Virgin of Izamal.

Returning to the main highway at Kantunil, turn toward Chichén Itzá. The flat countryside has, until now, been quite open, for the portion of the trip between Mérida and Kantunil–Izamal lies across another part of the heart of the henequen region seen at

Farther along, Libre Unión is noteworthy, at least locally, as the center of the socialist movement of the early part of this century—a movement mainly embodied by Felipe Carrillo Puerto, governor of Yucatán in the early 1920s who, with his brothers and others, fell before a Mérida firing squad in 1924. Carrillo's companion, the journalist Alma Reed, who is buried near Felipe in Mérida cemetery, inspired "La Peregrina," the quintessential traditional popular song of Yucatán.

The town of Pisté is famous in Maya archaeology as the source of skilled excavators, for the town continually furnished crews for the Carnegie Institution of Washington projects at Chichén Itzá and elsewhere for many decades. In recent years, Pisté has encouraged the building of tourist facilities—hotels, restaurants, and shops. You will note the results as we pass through. Chichén Itzá lies less than 1.2 miles (2 kilometers) away.

Chichén Itzá, a half day

Chichén Itzá is one of the most famous archaeological sites in the Americas and, at last tally, the most visited in all of Mexico. Yet it is also among the least known in terms of its historical chronology and ancient social makeup. These problems are due at least in part to conflicting accounts in the post-Conquest Maya chronicles, where history and prophecy are often inseparable, and to the emphasis on reconstruction and consolidation rather than on problem-oriented archaeology that characterizes most of the work done there since the early 1920s.

The Mayaland Hotel, built in 1930 by the Barbachano family, is one of the grand tourist landmarks of the area. Although it is in the midst of one of the most concentrated areas of ruins, the sprawling structure and its adjacent bungalows are well concealed by luxuriant tropical vegetation and gardens. From the lobby, the silhouette of the famed Caracol building is framed in the arch of the main doorway.

The ancient city—one of the very few such places in the Americas whose old name we know with certainty—thrived between about A.D. 700 and 1250, that is, in Late Classic and Early Post-classic times. In Yucatec Maya, still the local language, the name means the mouth of the Itzá well, the term Itzá referring to the family name of the rulers of the city during its heyday.

The conquistador of Yucatán, Francisco de Montejo, established himself briefly at the Chichén Itzá in the 1540s; two decades later, Bishop Diego de Landa visited the overgrown ruins and described the principal building and, at the end of the "handsome causeway," the now-famous Well of Sacrifice in his *Relación de las cosas de Yucatán*.

Landa's account was supplemented by dates and events re-corded in the Colonial Maya documents collectively known as the Books of Chilam Balam, named after a famed Indian seer. One—the Book of Chilam Balam from the town of Chumayel—deals at length with the role of Chichén Itzá in remembered history, and includes accounts (often conflicting among the sources) of the city's foundation, its shifting political alliances with the Cocom family of Mayapan and with Izamal, and the ultimate abandon-ment of the place early in the thirteenth century.

The site covers an estimated 8 square kilometers—its full limits have never been precisely determined—and its major axis runs north–south. The most readily accessible portion of Chichén Itzá is the northern part, where the hotels (including the Mayaland), the old hacienda, and the new Visitors' Center are.

Much of this northern area is defined by the huge artificial terrace that holds many of Chichén Itzá's most famous structures, including the Great Ball Court, the Temple of the Warriors, and the Group of the Thousand Columns—all dominated by the Castillo (castle), the most prominent pyramid. A causeway leads from it even farther north to the famed Well of Sacrifice, or Sacred Cenote.

Just south of the Castillo area—and due west of the Mayaland Hotel—lies another cluster of well-preserved or partially restored structures, which includes the High Priest's Grave, a small version of the Castillo; the Caracol, thought by some to have been an observatory; the Monjas (nunnery); and the Iglesia (church). (Most building labels, remember, exist mainly for convenience's sake and to our knowledge have absolutely nothing to do with the ancient function of the structures.)

The entire southern portion of the site is traditionally called Old Chichén. This label stems from the long-standing perception of Chichén Itzá as really being two sites in one—a Maya ruin, marked by Late Classic and Early Postclassic architectural style; the other, Mexican (specifically Toltec), distinct for its later Postclassic style.

This two-part concept of the site is useful, and indeed is reinforced by old chronicles that mention an invasion by foreign followers of Kukulcan, the Feathered Serpent, but may be a premature oversimplification of the true archaeological and cultural history of Chichén Itzá. Investigator Charles Lincoln finds little archaeological evidence that the place was ever anything other than Maya, and believes that the change apparent in the remains merely reflects a sort of Postclassic cosmopolitanism. Karl Taube thinks that, whatever the ethnicity of the city, there was a conscious effort to tie it to the standards of Toltec Mexico. Nevertheless, the site offers the best view available of Postclassic splendor in the Maya area.

Fitting well into the canons of Late Classic Maya architecture are structures that include the Monjas Complex, the Akab Dzib (House of Dark Writing), the Temple of the Three Lintels, and the Iglesia. These are all done in the Puuc style, with veneer masonry, and often feature intricate mosaic facades (the latter feature reaching perhaps its most extreme expression here in the Iglesia). All of these structures lie more or less in the southern reaches of the site.

Mexican, or Toltec, Chichén Itzá is manifest in the architectural decor that includes square columns carved with warrior figures, feathered serpents as columns and balustrade decoration, and the use of great colonnade–patio combinations in building plans.

The so-called Castillo is the epitome of Mexican-style architecture at Chichén Itzá, and clearly was the principal structure at the site. It is noted for its great columns made in the image of feathered rattlesnakes, and for the same motif that decorates the balustrades of the north staircase. Two of the four sides of the pyramid–platform have been restored; the others show the condition of the structure when it was first noted by the Spaniards and other early visitors. An almost identical structure with four staircases lies inside the Castillo pyramid, and may be seen in part by means of a tunnel whose entry is at ground level on one side of the north staircase of the visible building.

Most of these buildings in the north, in the vicinity of the Castillo, contain elaborate bas-relief sculpture, much of it still bearing its original paint. Of particular note are the panels of the Great Ball Court (Mesoamerica's largest ancient game field), the Tzompantli (skull-wall) and Venus platforms in the plaza north of the Castillo, and the interior walls of the Lower Temple of the Jaguars. Merle Greene Robertson, among many other things, studies surviving Maya pigments and paints, and is currently engaged in making rubbings of all the relief sculptures at this amazing site.

Free-standing sculptures in this same area include examples of the Chac Mool, the half-reclining figures—they are in the posture of one braced on elbows after waking up—that have evolved into sort of a trademark of the site. The name Chac Mool means nothing; it was invented by the eccentric Augustus Le Plongeon, who excavated at Chichén Itzá in the 1870s and who imagined it the name of an early ruler of the place.

The Temple of the Warriors, just east of the Castillo, is remarkable for its colonnades and carved square columns bearing the images of warriors. Its summit temple boasts feathered serpent columns and a well-preserved Chac Mool.

The Sacred Cenote, or Well of Sacrifice, lies at the end of a short causeway that extends northward from the area of the Castillo. The well, a great natural sinkhole in Yucatán's porous limestone, is nearly circular with an average diameter of about 200 feet (65 meters). A sheer vertical wall of vine-draped stone reaches from the surface to the water, some 70 feet (25 meters) below. The nearly stagnant pool is only a few feet deep along its edge; the center depth, about 40 feet (15 meters). The bottom consists of the ooze of the millennia, intermixed with fallen trees, stones, and waterlogged brambles. In and under this debris lay the treasure dredged up by Edward Thompson (b. 1860; d. 1935) between 1904

The Sacred Cenote (well) at Chichén Itzá was the site of human sacrifice. Photograph by Barbara Lerner Kopel.

and 1922—some 30,000 artifacts, including pottery, incense, gold and copper ornaments, and skeletons—mainly of children apparently sacrificed to the rain god. The National Geographic Society participated in another "excavation" of the Sacred Cenote in 1961, when Eusebio Dávalos Hurtado of Mexico's National Institute of Anthropology and History, and others, recovered most of what Thompson had left.

The Sacred Cenote continued to be used as a destination of pilgrimage and offering well after the abandonment of the city itself, around 1200.

The relatively new Visitors' Center at Chichén Itzá lies immediately southwest of the terrace containing the Castillo. It contains exhibits of sculpture from the site as well as a number of shops, both fancy and modest, a bookstore, a snack bar, and restrooms.

Detail from the Iglesia at Chichén Itzá depicting a mask of the rain god Chac. Photograph by Barbara Lerner Kopel.

In general, the "pure" Maya-style edifices south of the Castillo provide an architectural contrast to the Mexican-style buildings to the north. The Chicchan Chob (Red House) and the Akab Dzib (House of Dark Writing) are models of elegant simplicity, and both contain Maya hieroglyphic inscriptions.

The Caracol (Snail), the round building thought by some to have functioned as an observatory, is so-named for the spiral passageway that ascends through the interior core of the structure. The late Mayanist Eric Thompson, who likened the Caracol to a just-unwrapped wedding cake, thought it one of the ugliest buildings in the Maya area. Others disagree. Whatever one may think of the building, it is a rare and interesting type, and one that went through many stages of construction. It exhibits both Maya and Mexican traits, the latter reinforced by the presence of the stone incenseburners that line the final stage of the platform upon which the building rests.

SOUND AND LIGHT SHOW

A "light and sound" display is presented just after dark in the Castillo plaza area—where you may have seen the metal coverings over the buried light fixtures during the day. It involves sitting assembled near the south end of the Great Ball Court to hear a loudspeaker narration about the site and its history while the buildings are in turn bathed in colored light to illustrate the account.

The so-called Monjas is one of the largest structures at Chichén Itzá—a massive and multistoried complex containing many hieroglyphic lintels still in place over their doorways. Inside, murals—one depicted large boats—once graced some rooms, but most of these are gone.

The Iglesia, just off the northeast corner of the Monjas building and its annexes, is one of the most elaborate Maya buildings in existence, and it is virtually intact. Top-heavy to even the most casual observer, the Iglesia boasts an incredibly complicated mosaic facade dominated by great deity faces, including the long-nosed Chac, or rain god, masks that jut from the corners.

One of the best things about this area of the site is that it is a principal center for activity of the turquoise-crested motmot, one of the loveliest of the birds of the area. The best time to see these wondrous creatures is around sunrise, along the wooded trail behind the Monjas building. Depending on permission, inclination, time, and weather, you may wish to take the back trail behind the old hacienda (now a hotel) to the woods behind the Monjas for bird-watching.

◢ *Day 9*

COBÁ

Chichén Itzá to Cobá

The drive eastward from Chichén Itzá brings one almost immediately to the turn-off for the famed Balancanche Cave, discovered in 1959 and mapped and examined with National Geographic Society funds by E. Wyllys Andrews IV. Sealed chambers of the cave, revealed by Humberto Gomez, a guide from the Mayaland Hotel, contained Late Postclassic incense-burners and offerings of miniature metates (corn grinders). The area around the entrance to Balancanche has been converted into a delightful and informative botanical garden.

Soon after the Balancanche turn-off, a side road to the right marks the way to Chan Kom, the site of landmark studies by Carnegie Institution and University of Chicago ethnologists in the 1930s. A few kilometers farther on, the highway passes through the town of Kaua, known as the source of an important Colonial Period Maya chronicle.

The city of Valladolid (the ancient Maya city of Saki) boasts some interesting Colonial Period buildings on its main plaza, and a famous underground cenote, a few blocks away. The Valladolid of 1847 had become "the Sultaness of the East"—a city perceived by most observers as one of pretentious aristocrats located on the very frontier of essentially ungoverned Maya. For this and other reasons, Valladolid that year became the scene of a massacre by

Chichén Itzá to Cobá

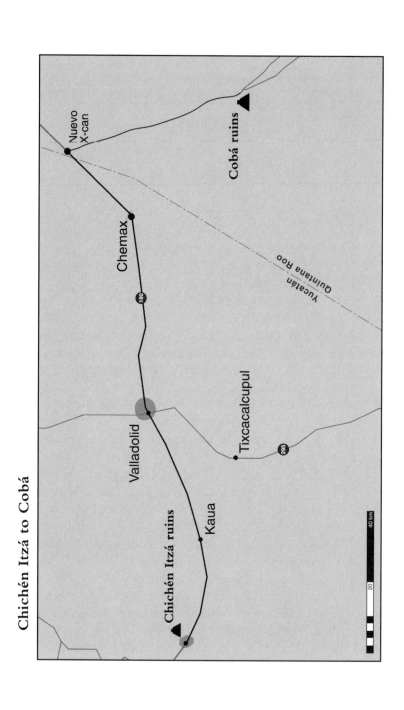

Maya troops—an incident that did much to set the pattern of the War of the Castes, which lasted until 1902.

About 15 miles (24 kilometers) north of Valladolid lie the ruins of Ek Balam, currently under investigation by William Ringle of Davidson College. There, Ringle's map has revealed an unusual pattern of building—concentric walls at the site center from which raised roadways go outward along the cardinal directions, perhaps a reflection of some effort to express Maya cosmology in the very pattern of the layout.

The town of Chemax—the highway bypasses it, but the seventeenth-century church may be seen over the dark line of bush—is notable as the nearest old town to Cobá village (a fair road between the two was recently completed), and as a seat of Maya rebellion during the Caste War of the nineteenth century. Here, one also begins to see old windmills (made by the Chicago Aero-Motor Company), remnants of thousands that once dotted the henequen zone of the northern Lowlands in such great numbers that they became a hallmark of the Yucatán landscape.

From Chemax, the highway begins to pass through an area that has only recently begun to be settled (and deforested). The settlements of X-Can and Nuevo X-Can lie near the old boundary between Quintana Roo territory and the state of Yucatán where, before 1974, one had to clear customs. At Nuevo X-Can lies the turn-off to Cobá, to the right.

The drive south from Nuevo X-Can to Cobá passes through country that is undergoing its first modern settlement, and the seemingly endless forest yields each day to new farmsteads, settlements, and access roads. At Punta Laguna (about 15 miles north of Cobá), an extensive Postclassic ruin with many complete buildings lies near a large lake.

Cobá

No buses go to this site, but tours can be arranged from Cancún and Playa del Carmen. The site, at least 50 square kilometers, is open from 8:00 a.m. to 5:00 p.m.; a small admission is charged.

Cobá

Uxulbeuuc

Grupo
Nohoch Mul

Grupo
Chumuc Mul

Grupo Cobá

Grupo
Magonxoc

Lab Mul

Lago Magonxoc

Uitzil
Mul

Domingo
Falcon

Lago Cobá

Pech
Mul

Chikin
Cobá

Dzib Mul

After: Catillo, A. B. Los caminos de Cobá y sus implicaciones sociales, 1981

375   750 m

To get to Cobá, turn off 0.6 mile (1 kilometer) south of the Tulum junction, which is 50 kilometers from the main road. Go northwest from Tulum on Hwy 307 to Chetumal; turn left and proceed 1.2 miles (2 kilometers); turn right and continue 26.1 miles (42 kilometers); from Tulum buses leave for Cobá at 6:00 a.m. and noon and take about an hour and a half.

The ruins of the ancient city of Cobá occupy an area of some 60 square kilometers. In general, the ruins cover an irregular ellipse of land oriented southwest to northeast, centered on the arcing chain of lakes. The center of Cobá is defined by several major groups of buildings and platforms, strategically arranged near Lakes Cobá and Macanxoc, the two largest of five shallow but permanent bodies of water that give the region its distinct character. The area is also the location of the modern town of Cobá, founded in the 1940s. The Villa Arqueologica was built there in the late 1970s, when the access roads from Tulum and Nuevo X-Can began to increase the influx of tourists to the site.

Although the official discovery of Cobá is attributed to the famed explorer-photographer Teoberto Maler, who visited the ruins in 1891, the first recorded visit by outsiders dates to the summer of 1886, when two travelers from Mérida, J. P. Contreras and D. Elizalde, made a brief visit to the overgrown ruins and deserted *chiclero* (harvester of the resin [chicle] from which chewing gum is made) shelters beside Lake Cobá. The Mayanist Thomas Gann (who thought he had discovered the place) described Cobá in 1926. The modern investigation of the site began at the same time with various trips by staff members of the Carnegie Institution of Washington who were then engaged in the exploration of nearby Chichén Itzá, culminating in the first major publication of the site in 1932. Since that time, important investigations of Cobá have taken place under the auspices of Mexico's National Institute of Anthropology and History. In 1974 and 1975, William J. Folan and George Stuart, in collaboration with Mexican archaeologists, mapped some 6,700 structures here.

The investigation of Cobá indicates at least two major periods of occupation between its beginnings in the Late Preclassic Period (ca. 300 B.C.) and its abandonment in Late Postclassic times (ca.

A.D. 1500). The major constructions of the three largest groups—and probably many of the outlying complexes as well—appear to date from the Classic Period, between about A.D. 500 and 800, a span defined by ceramic analysis, for most of the thirty or so monuments at the site are badly eroded. Among the exceptions to this is the well-preserved Stela 20, discovered in front of a vaulted building near the Nohoch Mul pyramid. It bears a clear Long-Count date of 9.17.10.0.0 12 Ahau 8 Pax, equitable (in the Thompson correlation) to 2 December A.D. 780—the latest date so far known from Classic Period Cobá.

At some point in the Postclassic Period, perhaps after a general abandonment of the site that may have lasted a century or two, Cobá was reoccupied, and its earlier, Classic Period constructions served as the foundation for a second major occupation. During this time, between about A.D. 1000 and 1500, the buildings and shrines of the Pinturas Group and many others, including the summit structure of the Nohoch Mul pyramid, were erected over much of the site. In addition, many of the fallen Classic Period stelae, even broken and eroded fragments, were reset in new locations. At some point, the whole place was abandoned, not to be resettled until the beginning of the modern town, some 400 years after the Spanish Conquest of the 1540s.

The uniqueness of Cobá derives from several factors. In size it is, along with Tikal and Calakmul, among the largest in all the Maya Lowlands. Its system of roadways, both internal and external, is not matched by any known Maya site. And, finally, the site itself lies virtually untouched by archaeology and still in a relatively intact natural setting, much as early explorers saw it.

The Cobá Group occupies most of the narrow bridge of land between Lakes Cobá and Macanxoc and extends along the entire northern shore of the latter—a total distance of some 2,296 feet (700 meters). Its eastern portion consists of a complicated series of quadrangular plazas enclosed by fallen structures; the western part features an immense plaza adjacent to the east end of Lake Cobá bordered by long mounds on the north and south. The elevated range of the plazas and buildings facing the plaza on the east side is dominated by the Iglesia, the tallest pyramid-temple of the

The Iglesia in the Cobá Group at Cobá. Photograph by Barbara Lerner Kopel.

group. Just north of it lie the ruins of a ball court, one of two known at the site. A gigantic system of ancient raised roadways gives Cobá perhaps its most distinctive characteristic. All in all, there are some fifty of these features at the site, of which the most notable is sacbe 1, which runs from the center of the site all the way to Yaxuná—a distance of 62 miles (100 kilometers)! Another, sacbe 16, leads 15 miles (24 kilometers) to the southwest to the small ruin of Ixil. And, like the spokes of a great wheel, others of these raised roadways of stone and gravel lead to the main groups that comprise "suburban" Cobá—Kucicán, Nuc Mul, Kitamná, Chan Mul, San Pedro, Telcox, and others. These vary between 2 and 6 kilometers (1.2 and 3.7 miles) in distance from the lakeside downtown groups and, in the general plan of the whole, form a sort of skeleton for greater Cobá. In the zones adjacent to or between these great roadways lie literally thousands of structures—clusters of house platforms, an occasional group of small vaulted build-

Palm leaf awnings are erected to protect ancient paintings on many structures such as this one in the Pinturas Group at Cobá. Photograph by Barbara Lerner Kopel.

ings, and the intricate network of low masonry walls that apparently served to define household areas and garden plots in ancient times.

Immediately southeast of Lake Macanxoc lies the slightly smaller Macanxoc Group, a series of large mounds and stela platforms on a large terrace. Macanxoc is most famous for two monuments upon which the Classic Period Maya thrice recorded an expanded Long-Count date expressed in twenty places above the normal *katun* position—in other words, the mathematicians of ancient Cobá produced a number equal to a number of years that we would have to express as 142 followed by thirty-six zeros!

Near where the trail turns off toward Macanxoc lies the Pinturas Group, one of the main Postclassic groups that occupy the Cobá zone. This group is dominated by a well-preserved pyramid, the summit temple of which still bears remnants of important murals depicting the maize god and a hieroglyphic text. The latter is the only inscription we know of outside the four codices that has survived from this late period. Here one may also see the remains of numerous altar-shrines and earlier, Classic Period monuments that were re-erected in Postclassic times.

The largest group of all at Cobá, the Nohoch Mul Group, is situated approximately 0.6 mile (1 kilometer) northeast of Lake Macanxoc. It is dominated by the massive pyramid-temple that gives the group its name (Nohoch Mul means large mound in Yucatec Maya). Nearly 98 feet (30 meters) high, this is the tallest single structure at the site, and the building on its summit—a three-doorway vaulted structure—is one of the best preserved. The Nohoch Mul faces south, toward a huge gravel plaza, itself enclosed by a series of ruined vaulted buildings. And just west of the great structure lies one of the largest artificial constructions in all of the Maya area, a colossal four-story complex of vaulted buildings and stairways with a flat summit. It measures about 361 by 410 feet (110 by 125 meters) at its base and features a grand staircase on its southern side.

Other important groups at Cobá include Group D, off in the woods between the Pinturas Group and the Nohoch Mul; the Chumuk Mul Group, between the Cobá Group and the Nohoch Mul; and the various architectural clusters that lie at the ends of the radial system of causeways, or sacbe, that originate in downtown Cobá.

In its heyday(s), Cobá apparently served as a powerful regional center and trade capital, perhaps a commercial hub in the distribution and redistribution of goods between the interior of the northern Lowlands and the east coast ports. It also clearly served as a seat of powerful rulers and as such doubtless played a key role in the rivalries with neighboring states such as that centered at Chichén Itzá, to the west and north.

Cobá to Tulum, to Cancún

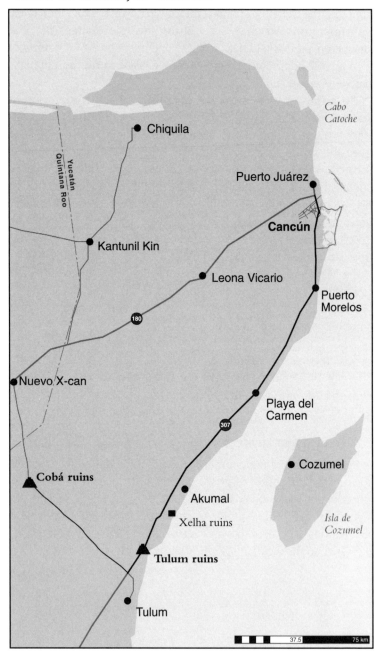

△ Day 10

CANCÚN

Cobá to Cancún

The drive from Cobá to Cancún follows the road to Tulum and the east coast, begun in the late 1960s but unfinished until the mid-1970s. Here, too, the highway is occasionally punctuated by new settlements, and new electric power lines. About 25 miles (40 kilometers) out of Cobá, the paved road intercepts the main coast highway. To the right lies the town of Felipe Carrillo Puerto and points south; to the left, the ruins of Tulum and, eventually, the resort of Cancún.

The ancient walled coastal city of Tulum, accessible by a short access road on the right as one proceeds north, features many buildings and some extraordinarily preserved murals. This site, two hours by bus from Cancún, bears the full brunt of tourism, and this is manifest in the market stalls of the expected clothing, crafts, and other goods that, each year, seem to fill more and more of the large parking lot just outside the old main gateway in the wall, competing for space with the parked files of idling buses.

The four main ruins at Tulum can be visited in about an hour. The Temple of the Frescoes, nearest the road, is a two-story, columned structure that contains the frescoes for which Tulum is known. They depict the worlds of the ancient Maya and their major deities. Nearer the sea in conjunction with two smaller structures is the Castillo, from the top of which you look down 40 feet (12 meters) to the sea below. (North of the Castillo, a path

Tulum

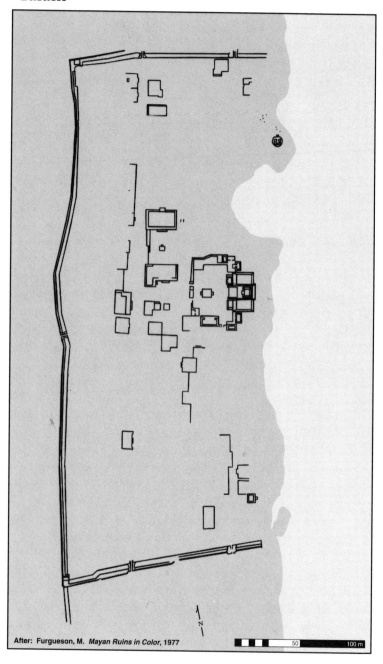

After: Furgueson, M. *Mayan Ruins in Color*, 1977

N

50   100 m

Tulum is an enchanting site because of its dramatic location on a cliff overlooking the Caribbean Sea. Photograph by Barbara Lerner Kopel.

down to the sea has been carved into the cliff.) The Temple of the Descending God, named for the apparently descending figure carved above the doorway, is north of the Castillo; and the Gran Palacio is northwest of that.

At Tancah, a few kilometers north of Tulum, the highway passes beside a nearly intact Maya building on the right, or coast side, of the road. It was here, and at the many other structures in the surrounding bush, that Arthur Miller found and recorded many important Late Postclassic murals. Needless to say, those inside the building beside the highway have gradually yielded to the vandalism of passers-by and the mold of time.

Beyond Tancah, Xclha has ruins, visible on both sides of the highway, and a series of crystal-clear, brackish lagoons that contain an astonishing variety of tropical fish. There is a museum on the site, now a regional park, and swimming is permitted in some of the lagoons.

The main portion of the ruins of Xelha lie on the side of the highway away from the coast. Two main groups of buildings are open here. One features some well-preserved murals; the other, some well-preserved buildings in quiet forest near a deep cenote (natural sinkhole well).

Akumal, another resort-cum-ruins, is best known as a resort—one of the oldest on the coast—where CEDAM, the Mexican diving club, has its headquarters. There is a small museum at Akumal, devoted largely to artifacts found by members of the club.

That portion of the road between Akumal and Cancún passes near Xcaret and numerous other places that have gained at least minor fame for their archaeological remains, as small-scale resort retreats, or for their combinations of these features. Two rapidly growing towns, Playa del Carmen and Puerto Morales, both terminus points for the ferry to Cozumel, are also on this highway, completed in 1969, while Quintana Roo was still a territory. (Quintana Roo achieved statehood in 1974, when Cancún began to look like a formidable reality.)

Xcaret (or Pole), not visible from the highway, is currently under investigation by archaeologists from Mexico's National Institute of Anthropology and History. Xcaret, like most of these east coast ruins, contains numerous well-preserved buildings, some abnormally small, that date from the Late Postclassic Period.

Cancún

Cancún (population, 115,000) is the terminal for buses from Mérida, Chetumal, and Mexico City. From here you can fly to Cozumel or take a bus to Playa del Carmen (43 miles, or 70 kilometers) and then take the ferry (50 minutes) to Cozumel. Tulum is two hours by bus from Cancún.

Cancún is a contrived resort, born of a deliberate computer search in the 1960s, and hastily but fairly well built through the years since. Its steady growth was briefly interrupted by the double-barreled devastation of hurricanes Gilbert and Keith in 1988. Before any construction started, archaeologists and other visitors knew Cancún as a wide and virtually featureless beach fronting a turquoise sea, with no human works in sight. The only break in the beach line at the time was a large tilted formation of dark limestone that stuck out like the prow of a ship gently sinking. On its wide flat summit lay a pristine archaeological midden dating from pre-ceramic times, or before about 2000 B.C. That site, briefly investigated by E. Wyllys Andrews IV, now fronts the Chac Mool Hotel, and goes generally unnoticed, though a hard and deliberate search may still reveal fragments of stone tools, shell, and even a piece of human bone or a tooth now and then.

As you can see on the map, Cancún consists of three main parts—the city, the international airport, and the hotel area. The city grew rapidly with the resort, from a tarpaper shantytown—a base for the construction workers—to its present state (still growing, but a bit slower than before)—a boom town with much of the character of a border town.

To separate the city of Cancún—which, after all, has little to offer the majority of those who come here—from the flow of paying visitors between the beach hotels and the airport, the planners placed the latter far to the south of Cancún city, at a spot where direct access to the beach area would be easiest. Thus, tourists proceed directly to the hotel area via its southern end without getting near Cancún city.

The hotel area is built on the great sand spit—it measures about 10 miles (16 kilometers) north–south—that encloses a large lagoon. The "best" hotels, many of them architectural wonders, and nearly all replete with fashionable shops and expensive restaurants, now line nearly the whole of the north and east portions of the beach area. The southern end of the beach area, nearest the airport, is the least developed area of all so far. The entire length of the hotel area is linked by the Avenida Kukulcan, a four-lane highway usually filled with fast-moving taxis. Here and there on the grassy median strip are set concrete replicas of famous monuments of Mexican antiquity, including a colossal Olmec head and various Aztec sculptures. Shopping malls are increasing in number almost every year, with a general concentration near the northeast corner of the hotel area. The lagoon enclosed by this incredible strip is now lined by condominiums that have grown up on its mainland edge. One bright spot on the Cancún conservation front: Many of the hotels have set aside special areas of beach so that the sea turtles can come ashore unmolested to lay their eggs.

There are two principal archaeological sites in the Cancún hotel area, but if you miss them, little will be lost. One, now represented by a ruined Postclassic pyramid with summit temple remains, lies next-door to the Sheraton Hotel. It affords a good view of sunrise over the ocean. The other, called El Rey, lies on the lagoon side. At the latter, another Postclassic site, many building platforms, walls, and columns have been excavated and consolidated. And if you want to see lots of sleepy iguanas, visit El Rey!

The area around Cancún holds much of interest. Just to the north of the hotel area—and visible on the horizon—lies Isla Mujeres, a relatively undeveloped and quiet resort named by the Spaniards who came to this area in the early 1500s.

The mainland from just north of Cancún up to Cape Catoche is a sort of archaeological terra incognita, save for the ruins of El Meco, near Puerto Juárez, and the ruins of the large sixteenth-century church that may mark the site of Ecab, spotted in 1517 by the Córdoba voyage credited with the 1517 discovery of Yucatán.

Recent analysis of satellite images by Charles Duller of NASA's Jet Propulsion Lab has revealed what may be ancient canals in the area as well.

Reaching Cancún is a milestone of sorts, for in doing so, the visitor has come full circle and ended up on the very shore where the Spaniards first touched the astonishing world of Mesoamerica.

PART THREE

Resources

△ Hints to the Traveler

☐ Citizens of the United States and Canada don't need a passport, but it would probably be a good idea to take one. What you do need is a tourist visa (*pasaporte*), which is good for 180 days for U.S. citizens; 90 days for Canadians. You need proof of citizenship (e.g., a driver's license). If you arrive on commercial airlines or ship, the carrier will do the paperwork. The important point is to remember to get it stamped when you arrive and to return it when you leave.

☐ Mexico is deservedly known for its beer. There are about thirty brands, many with local distribution (e.g., Leon Negro is brewed and distributed in Mérida).

☐ Beachwear is never worn on the streets in other than a tourist resort; and bare feet are not acceptable on the street unless you are destitute.

☐ Public transportation is generally good.

☐ The country is mountainous. There are many air routes, but air travel is six or eight times more expensive than bus.

☐ Taxis are available in even the smallest villages. Agree on the fare before entering the cab! Generally cabs are a bargain.

☐ Taking your car into Mexico is difficult and expensive. Only "Mexican Auto Insurance," issued by local underwriters, is recognized. Auto accidents are handled under the criminal code; therefore an uninsured driver could spend six or seven months in jail waiting for the case to come to court.

☐ Car rental are generally expensive. Train fares are lower than bus fares but also much slower: "on time" is the exception.

☐ The rainy season is May to October.

□ When visiting the ruins, be sure to take a hat, sunscreen, and water. Arrive at the ruins as early as possible to avoid the crush of the tour buses, which get there around midday.

□ It's helpful to know some Spanish, but not essential. The more remote the area, the fewer people who will speak English. In the most remote places, the people speak only their own language and not Spanish.

△ Suggested Readings

Adams, Richard E. W.
1991 *Prehistoric Mesoamerica*. Norman: University of Oklahoma Press.

Coe, Michael D.
1988 *The Maya*. London and New York: Thames & Hudson. Fourth edition.

Flannery, Kent V.
1983 *The Cloud People*. New York: Academic Press.

Hammond, Norman
1988 *Ancient Maya Civilization*. New Brunswick, NJ: Rutgers University Press.

Kelly, Joyce
1982 *The Complete Visitor's Guide to Mesoamerican Ruins*. Norman: University of Oklahoma Press.

Kowalski, Jeff
1987 *The House of the Governor: A Maya Palace at Uxmal*. Norman: University of Oklahoma Press.

National Geographic
1986 Supplement Map "Central America, Past and Present." *National Geographic*, April.

Reed, Nelson
1964 *The Caste War in Yucatán*. Stanford: Stanford University Press.

Roys, Ralph
1957 *Political Geography of the Yucatán Maya*. Washington, DC: Carnegie Institution of Washington.

Sabloff, Jeremy A.
1990 *The New Archaeology and the Ancient Maya*. New York: W. H. Freeman and Co.

Schele, Linda, and David Freidel
 1990 *A Forest of Kings.* New York: William Morrow.
Sharer, Robert J., Sylvanus G. Morley, and George Brainerd
 1986 *The Ancient Maya.* Stanford: Stanford University Press. Fourth
 edition of the original by Morley.
Stuart, David, and Stephen D. Houston
 1989 "Maya Writing." *Scientific American,* August issue.
Stuart, Gene S.
 1981 *The Mighty Aztecs.* Washington, DC: The National Geographic
 Society.
Stuart, George E., and Gene S. Stuart
 1977 *The Mysterious Maya.* Washington, DC: The National Geographic Society.
Winter, Marcus
 1989 *Oaxaca: The Archaeological Record.* Mexico: Minutiae Mexic-
 ana.

△ Index

Abaj Takalik, 15, 16
Acancéh, 87
agriculture, 8
Aguilar, Gerónimo de, 44
Akumal, 112
Alameda Park, 52–53
Alvarado, Pedro de, 43
Archaic Period, 7–8
Azcatpotzalco, 31
Aztec, 32, 35, 55

Balancanche Cave, 99
Becerra Formation, 6
Bonampak, 27, 42
Books of Chilam Balam, 93
Brasseur de Bourbourg, 45

cacao, 16
Calakmul, 22, 27
calendar, *see* Long Count
Can Ek, 43
Cancún, 42, 109, 112–114
Caracol, 27, 28
Carillo Puerto, Felipe, 92
Catherwood, Frederick, 86
cenote, 95–96
Cerros, 22
Chaan-Muan, 42
Chac, 35, 36, 40, 97
Chalcatzingo, 11
Chan-Bahlum, 74
Chan Kom, 99
Chemax, 101
Chenes, 35

Chiapa de Corzo, 16
Chiapas, 7, 46
Chicanel tombs, 22
Chichén Itzá, 33–35, 37–39, 88–90,
  92–100
Chol, 74
Cholula, 31
chronology, 21. *See also* Long Count
Chumayel, 87, 93
Classic Period, 23–31, 104
clothing, 79, 89, 91
Clovis, 6
Cobá, 27, 41, 80, 99–109
Colonial Period, 43–46
colonias, 52
Columbus, Bartholomew, 41
Columbus, Christopher, 39
Conquest, 43, 45
Copalillo, 10–11
Cortés, Hernán, 32, 43, 44, 51,
  62
Covarrubias, Miguel, 10
Cozumel, 44, 112
Cruzob, 46
Cuello, 20
Cueva Blanca, 6
Cuicuilco, 14, 54

Dainzú, 12, 66–67
Díaz, Porfirio, 62
Díaz del Castillo, Bernal, 44
Dos Pilas, 28, 31
dwellings, 9, 28, 39, 45
Dzibilchaltún, 79